The Economics of the European Patent System

D1261080

The Economics of the European Patent System

IP Policy for Innovation and Competition

Dominique Guellec

and

Bruno van Pottelsberghe de la Potterie

OXFORD
UNIVERSITY PRESS

Great Clarendon Street, Oxford OX2 6DP

Oxford University Press is a department of the University of Oxford.
It furthers the University's objective of excellence in research, scholarship,
and education by publishing worldwide in

Oxford New York

Auckland Cape Town Dar es Salaam Hong Kong Karachi
Kuala Lumpur Madrid Melbourne Mexico City Nairobi
New Delhi Shanghai Taipei Toronto

With offices in

Argentina Austria Brazil Chile Czech Republic France Greece
Guatemala Hungary Italy Japan Poland Portugal Singapore
South Korea Switzerland Thailand Turkey Ukraine Vietnam

Oxford is a registered trade mark of Oxford University Press
in the UK and in certain other countries

Published in the United States
by Oxford University Press Inc., New York

British Library Cataloguing in Publication Data

Data available

Library of Congress Cataloging in Publication Data

Data available

Typeset by Newgen Imaging Systems (P) Ltd., Chennai, India
Printed in Great Britain
on acid-free paper by
Biddles Ltd., King's Lynn, Norfolk

ISBN 978–0–19–929206–6
ISBN 978–0–19–921698–7 (Pbk.)

10 9 8 7 6 5 4 3 2 1

We dedicate this book to our closest support in all this
endeavour: Véronique, Clément, Lou, and Elena.

■ CONTENTS

■ LIST OF FIGURES, TABLES, AND BOXES

FIGURES

TABLES

BOXES

■ FOREWORD

I was both flattered and pleased when the authors asked if I would write the foreword to this book. I was flattered by an invitation from two energetic economists with whom I have worked with great interest. I was pleased because I believe passionately that we need to promote deeper understanding of the patent system and debate about what is actually happening, and what should happen next. This book is a vigorous contribution to that debate.

For long the uncontentious territory of the technocrat and the lawyer, the patent system has in fact a turbulent and disputed history and the twenty-first century is seeing the resurgence of some of the furious debates of the nineteenth century. One school of thought would say that this is simply noise, and not very well informed noise at that. But it is clear that identifying 'the right approach' to the patentability of computer implemented inventions, and indeed biotechnology, provokes very varied views in Europe, as it does elsewhere, and this is not simply noise.

The authors describe this as a handbook on the economics of patents: the concept is clearly every bookseller's dream. But my sense on reading it is that it is much more than a handbook on economics, and offers more provocation than one might expect from the title. I also note that the authors say that the patent system in Europe has lost its sense of mission, not a normal economic term I suggest, but the sense is vivid. The 'what for?' of the patent system has indeed become obscure (at least to some) and the 'where next?' is also unclear.

This book is about Europe: rightly so. The US constitution, with its belief in self-evident truths, reveals the certainties of the rational mind of the Enlightenment. The founding fathers in Philadelphia made provision for patents, though Jefferson was an early dissenter. Now much of the interesting debate about what the real economic effects of the patent system are, and whither it should go, comes to Europe borne on transatlantic wings. This is stimulating, but we need to think for ourselves too. This book will help us to do so. Its ideas make it more than a handbook. They will not find universal favour (some may be over-ambitious, or over-rational). But the push to consider more effectively what we should be aiming for and could achieve is very clear, and entirely welcome.

Patents are powerful economic tools (and I greatly welcome the references in this book to competition policy in the context of intellectual property). Powerful tools may become commonplace, but that does not mean that it is sensible or safe to treat them casually. I hope that this book will help us in Europe to come to a better understanding of what we are trying to do with patent policy, and how best to do it.

(Alison Brimelow, President-elect of the EPO, Great Malvern)

■ PREFACE

I accepted to write the preface to this book with great enthusiasm. To the best of my knowledge it is the first time I have seen a book that comprehensively tackles the various 'economic' dimensions of a patent system: from its history and its profound theoretical roots to the challenges it currently faces.

The EPO is a worldwide centre of excellence for patent searches and examinations. It is probably the patent office that ensures the highest quality in this kind of service, and offers a strong jurisdictional security to the patentees. The European patent system is also a network of national patent offices which effectively collaborate through the EPO. This diversity, which typically reflects the construction of Europe, ensures a convergence towards a more harmonized patent system in Europe.

For outsiders, and for its users, one of the main drawbacks of the European patent system is in the fact that the enforcement of patents takes place at a national level. The diversity of national jurisdictions induces a substantial level of uncertainty in this respect. An important step forward may be achieved through the currently heavily debated establishment of a centralized European court for patent litigations.

The patent granting authority aims at fostering innovation and competitiveness. As any system or institution, the EPO, which already contributes significantly to the innovation process in Europe, may be improved, and the authors provide several avenues for further brainstorming in this respect. If the benefits of a centralized court are clearly underlined, other possible directions for improvement are suggested by the authors, from a more in-depth consideration of the economic dimension of a patent system, to an even more rigorous and timely examination process.

From my point of view, beside the comprehensive analytical approach taken by the authors, beside their creative vision of the patent system, this book is warmly welcome, as it provides the economic roots of a patent system to lawyers, inventors, and patent examiners. The legal and systemic insights to economists and industry managers involved is addressed as well, enabling them to better manage the economical risk related to patents. I am convinced that the readers are certain to better understand, and to endorse, what the European patent system aims at doing, so that better informed policy debates can take place in the future.

(Professor Alain Pompidou,
President of the EPO, Munich)

■ AUTHORS' PREFACE

The authors of this book have both had experience as chief economist of the European Patent Office (EPO). When reporting that position to fellow economists, many would ask, 'Why does a patent office need a chief economist?' When reporting it to many EPO colleagues we would be faced with the same question. This question simply reflects the difficulty two separate worlds have in understanding each other. In a way, this book provides an answer. It aims to tell economists, on the one hand, how the patent system is important to innovation and how it is a sophisticated construction and, on the other, to tell patent professionals how economics is core to their field and how their system cannot live without a mission. This is not straightforward.

Economists would not deny the importance of patents, but they tend to have a rough idea of the system, seen simply as a type of monopoly right granted to an inventor in order to allow him or her to recoup research costs and act as an incentive. This is true, this is the inescapable starting point, but it misses the richness of the field. One cannot understand the patent system without taking on board the highly sophisticated way it operates as a result of laws, institutions, and practice. One cannot offer proposals for operational reform or make oneself heard by patent policy makers, without basing one's suggestions on the complex reality of the system, which goes far beyond basic economic principles.

Practitioners, notably lawyers, tend to focus on the details of the system while overlooking, sometimes even denying, its purpose. Quite a few lawyers negate the legitimacy of the very notion of 'patent policy': patents are about rights, not about policy. This approach results in an inward looking and conservative attitude, denying the legitimacy of outsiders to question the system. The patent system cannot be properly understood without a clear reference to its (essentially economic) mission and the conditions (economic, political, technological, and legal) in which it exercises it.

The benefits to be obtained from joining economics and patent expertise are hardly news. We met quite a few patent practitioners, notably at the EPO, who had embraced this approach, and even promoted it forcefully. We hope this book will help them in their future work. This book is aimed both at introducing economists to the sophistication of the patent law and practice, and to introducing lawyers and practitioners to the economic mission of their field. It is centred on Europe as it is the region we know the best, and Europe contrary to the US, has been subject to few such studies.

We would like to express our gratitude to many EPO colleagues who introduced us, economists, to the beauty of the mechanics of the patent system, helping us through valuable discussions, reading parts of the book, or providing us with the proper conditions for writing it: Eugenio Archontopoulos, Alison Brimelow, Manuel Desantes, Wolfram Förster, Yves Grandjean, Christoph Laub, Ciaran Mc Ginley, Marc Pacci, Alain Pompidou, and Berthold Rutz.

Patent managers from the business world granted us the privilege of sharing their experience with us, notably Francis Hagel and Thierry Sueur.

We also want to thank fellow economists with whom we were able to share ideas and experience over the years: Paul David, David Encaoua, Bronwyn Hall, Dietmar Harhoff, Jacques Mairesse, Catalina Martinez, and Suzanne Scotchmer.

Last but not least, we are grateful to two colleagues, Nicolas van Zeebroeck and Niels Stevnsborg, who have co-authored with us two sections of this book: Chapter 4 and Chapter 6, respectively.

We must of course mention the usual caveat: 'The views expressed in this book are purely those of the authors and may not in any circumstances be regarded as stating an official position of the EPO, the OECD, or of the ULB'. More fundamentally, we remain responsible for all the limitations and possible mistakes of the book.

(Dominique Guellec and Bruno van Pottelsberghe,
Paris and Munich)

1 Introduction

Simply put, the patent system enables inventions to be a part of capitalism.

(Nathan Myhrvold, Testimony, United States Senate
Committee on the Judiciary, 23 May 2006)

1.1. Why an Economic Approach to Patents?

The economic role of patents is probably more important than ever before—and still patent systems are only marginally influenced by economists. This is especially true in Europe. When it comes to discussing patent law, case law and regulatory practice, legal scholars have by far the greatest influence and they are doing a good job overall, even from an economist' perspective. In fact, much of patent law and practice is already influenced by economic considerations, if not by economists. This is as a result of common wisdom and lessons taken from opposing economic arguments exchanged in courts and modelling case law. It is certainly not the case either that economists have enough understanding of the patent system for optimizing it on their own. However it is our view, and it is the raison d'être of this book, that the European patent system would benefit greatly from being more explicitly based on economic considerations. In times of rapid change, when an institution must adapt rapidly, it has to mobilize all relevant knowledge and evidence that can accelerate and enlighten its own change.

The central role of patents in the knowledge economy is illustrated by simple facts and figures. The number of patent applications at the European Patent Office (EPO) was more than 200,000 in 2006, a 150 percent increase since 1995. A similar trend was experienced in the US and most other countries. Patent offices all around the world are flooded with applications that far exceed their processing capacity. The amount of licensing contracts (of which a major part is backed on patents) is higher than US$100 billion worldwide. Patent matters frequently make the headlines of newspapers in Europe, including the discussions prompted by the European directive on software patents, the practice of the EPO in the field of biotechnology or the difficulties involved in setting up a Community Patent. In the US, high profile litigation cases are common, often involving hundreds of millions of dollars, and several a year going as far as the Supreme Court. In a less visible, but equally important way, most of the new products which fascinate consumers and on which they spend their money are full of patented inventions—be they drugs or mobile phones.

The feeling is widespread that the patent system as we know it, despite its impressive achievements, will need significant transformations in order to address the new challenges raised by the knowledge economy. Patents and other types of intellectual property rights (IPRs) (notably copyright and trademarks) were designed at a time when the production and dissemination of knowledge were small scale activities compared with tangible, manufacturing activities. They are now the very basis of growth and competitiveness; they are performed on a large scale and at a worldwide level; and they follow industrial logic instead of the previously dominant craftsmanship approach. The development of information technology has been instrumental in that change, and has spread to the whole economic engine. IPRs are one of the institutional tools which allow market and even non-market forces to operate for informational goods. If one wants market forces and non-market forces to be still effective in the knowledge economy, IPRs will require significant changes.

IPR policy is essentially about balancing the two opposite requirements of access (by the public) and appropriation (by the inventor). Rethinking this balance in various areas, and revising the ways access and appropriation are defined and implemented, are the core tasks of reformers. That is the point where, in our view, economists can bring an essential contribution to ensuring that the necessary legal principles are adapted to new economic and social realities.

The US is addressing these issues already. Government reports, by the Federal Trade Commission (2003) and by National Academy of Sciences (NAS 2004) have started bringing together economists and lawyers. High quality books and handbooks by economists and legal scholars (e.g., Jaffe and Lerner 2004; Scotchmer 2004) have helped develop the profile and intellectual quality of the patent debate in the US. Nothing comparable has been done in Europe. And if Europeans lag behind in these debates, they will end up having to adopt solutions found on the other side of the Atlantic—which might fit US specificities and interests but not necessarily those of Europe. It is all the more urgent for Europe to start this debate as the European patent system is at a crossroads.

The feeling of a crisis is widespread in the European 'patent community'. Most recent initiatives for renovating the system have failed—the Community Patent and the software patent directive. The debate between patent offices in Europe has centred on how to share the work and revenue among them in order to optimize their budgets and other constraints, while the mission of the patent system vis-à-vis society was essentially overlooked. The 'Lisbon objectives' set by EU heads of state in 2000, which aimed at establishing Europe as the most dynamic, knowledge-based economy in the world by the year 2010, will be missed by most European countries. Certainly the reasons for this failure are broad—related to the lack of structural reform in general and to the weakness of the European scientific and technological policy—but the near absence of patents from the 'Lisbon debate' is striking, as well as the absence of references (other than purely formal) to the Lisbon agenda in current patent debates. It seems that the patent system in Europe has lost its sense of mission, and that this mission is ignored by policy makers in charge of innovation. This is clearly not a good situation.

It is not the goal of this book to analyse this context and prescribe solutions. More modestly, we aim at laying the ground for future substantive debates by establishing bridges between the various parties involved: sharing key notions and knowledge, and improving mutual understanding. The patent system will not change before legal scholars, who shape it, have acknowledged the economic dimension of their activity, and before economists have adapted their reasoning to the specific institutional and legal context of patents. We aim to do this within the specific context of Europe, which differs substantially from the US context. Hence, our book is a handbook on the economics of patents, with an emphasis on the European context, and presents a mix of economic theory, legal features, and empirical illustrations.

What does differentiate an economic approach from a legal or a technical one? The legal approach addresses issues of fairness and balance of rights, of internal consistency of the system, and of consistency of patent law with other bodies of law. All this is necessary, as a good patent system should be internally consistent, articulated on strict and clearly defined rules. The patent system must be based on law to be effective—the alternative, a discretionary system, would go against its mission of enhancing market incentives to innovate. The legal approach is, however, incomplete as it does not address the missions of the system, its rationale: Why do we have patents in the first place? Which features in the system improve or reduce its ability to fulfil its role? If one does not start from these questions, it is impossible to really evaluate and improve the system as evaluation is about confronting the aims with the actual results and improvement refers to goals. And the aim of the patent system is economic in nature: it is to foster innovation and growth. Referring to the missions of the system is not necessary when it is working well—Why look at the map when you are not lost? But at a time when change is needed, one needs a map, a compass, to indicate the directions to take and those to avoid.

The economic approach is utilitarian: it refers to the benefits of society, to costs and benefits; it does not see patents as a 'natural right' that the inventor should have, but as a policy instrument which should be adapted by government in the interests of society. Patents aim at supplementing market forces, which do not lead on their own to the socially desirable level of innovation. The benefits of the patent system reflect its mission—to encourage invention and diffusion of technology; and its costs reflect its modus operandi, which is to restrict the use of inventions. To the legal notion of 'balancing rights', economists prefer 'balancing interests' and 'optimizing total outcome'. Not only the mission, but also the modus operandi of patents is economic in nature. Patents are economic incentives (increasing profits through market exclusivity) for reaching an economic goal (enhancing innovation and growth). Hence if one wants to improve the efficiency of the system, it needs to be referred to the way it works: what strategies do patent applicants and holders follow; how they interact with the system; how they use their patents; etc. Without a proper understanding of these issues by all parties involved, the public discussion will keep focusing on narrower questions and the patent system will not be in a position to evolve so as to better fulfil its mission.

On the other hand, economists and economic policy makers must acknowledge that the patent system must be based on law, not on discretion, and that its plasticity is limited, for good reasons. The system works by raising the expectation of market exclusivity for inventors. Inventors must then be in a position to make a reasonable and reliable expectation prior to engaging into research. That requires clear and stable rules. This is the limit to economic optimization: in the face of variegated economic circumstances, which call of variegated policies, the patent system requires some (although not absolute) uniformity. This constraint does not necessarily fit well with the ad hoc reasoning of standard economics; trying to escape from it to create a patent system based on discretion instead of law would mean breaking the very core of the system, which is certainly not desirable either.

1.2. **A Primer on Patents**

Before entering into more detailed analysis, it may be useful for non-expert readers to acquire some familiarity with the basic legal and institutional machinery of patenting and its vocabulary. They will find it in this section, which can easily be skipped by those already familiar with the subject.

A patent is a legal title protecting an invention, which gives the following rights to its holder:

1. A patent shall confer on its owner the following exclusive rights:
    (a) where the subject matter of a patent is a product, to prevent third parties not having the owner's consent from the acts of: making, using, offering for sale, selling, or importing for these purposes that product;
    (b) where the subject matter of a patent is a process, to prevent third parties not having the owner's consent from the act of using the process, and from the acts of: using, offering for sale, selling, or importing for these purposes at least the product obtained directly by that process.

2. Patent owners shall also have the right to assign, or transfer by succession, the patent and to conclude licensing contracts.

(Article 28 of the TRIPs)

Patents are territorial rights: they only apply to the jurisdiction for which they have been granted. Patents are granted to inventions from all fields of technology. Aesthetic creations, scientific theories, natural phenomena, or abstract ideas are not patentable.

Patents are obtained by following particular administrative procedures. In order to obtain a patent, the inventor has to file an application to a patent office, which is a government body that checks that inventions fulfil certain legal criteria, who will grant a patent or reject the application accordingly. There are different alternative procedures available in Europe, and the applicant will choose according to his or her national, European, or worldwide strategy (see Chapter 6 for a more comprehensive description of the patent process).

General Procedure

When an inventor (an individual, company, public body, university, non-profit-making organization) decides to protect an invention, the first step is to file an application with a national patent office (generally the national office of the applicant's country). The first application filed worldwide (in any patent office) for a given invention is known as the '*priority application*', to which is associated a '*priority date*'. The patent office then begins 'searching and examining' the application in order to check whether a patent may be granted or not—in other words, whether the invention falls into the patent subject matter and is novel, inventive ('non-obvious to persons skilled in the art'), and capable of industrial application. The application is published 18 months after it is filed ('*publication date*'). The lag between filing and grant or refusal of patents is not fixed, it ranges from two to eight years, with significant differences across jurisdictions.

International Application ('Paris Route')

Since 1883, when procedures were standardized under the Paris Convention (which has about 170 signatory countries in 2006), applicants who wish to protect their invention in more than one country have 12 months from the priority date to file applications in other Convention countries or not, and if they do so the protection will apply from the priority date onwards.

EPO (European Patent Office)

The EPO is a regional office which searches and examines patent applications on behalf of European countries, with 31 members in 2006. The EPO has been created by and implements the European Patent Convention (EPC), signed in 1973 in Munich, which fixes patenting procedures and criteria for Europe. The EPO grants 'European patents', which are valid in all its member states in which the holder has validated his or her right, and which necessitate translation into the national language and payment of national fees. On this national stage, European patents are submitted to national laws. On average European patents are validated in six to eight countries. In 2006, the EPO does not substitute entirely for national offices, which still search and examine national patent applications. The EPO has more than 6000 staff, including about 4000 examiners. Applications can be filed directly to the EPO; they can come after national priority; or they can be filed through the PCT route. In the first two cases they are labelled 'Euro-direct' applications.

PCT Procedure

Another procedure for protecting a patent in several countries is to file an application under the Patent Co-operation Treaty (PCT), which has been in force since the late 1970s and is monitored by the World Intellectual Property Organisation (WIPO). The PCT procedure is an intermediate step between the priority application and the filing for patent protection abroad. It extends the potential protection given by the priority right to a period of 30 months. During that time in the procedure the applicant has to decide in which countries (among the nearly 130 signatory states in 2006) he or she will exercise or not his or her rights. The application goes from the 'international phase' to the 'national phase' in the chosen countries, where it follows the standard national procedure, while it is abandoned in others. The PCT is a way of keeping open the option to file future applications abroad, it is not an actual patent application. The PCT application is searched by one of the *International Search Authorities* (ISA), patent offices labelled by the WIPO, the one chosen by the applicant. If the applicant designates the EPO as its ISA (the case in more than 50 percent of PCT applications in 2006), the application is known as a '*Euro-PCT*' application. Due to its convenience and cost efficiency, the PCT procedure has met with much success with patent applicants, and now more than two-thirds of applications processed by the EPO are PCT filings.

National patent laws have to comply with international standards, made in TRIPs (trade related aspects of intellectual property rights), an international treaty which is part of the WTO package signed in 1994. TRIPs imposes strict conditions on WTO members, such as patentability of all fields of technology, minimal duration of patents of 20 years, limitations of compulsory licensing, etc.

After its grant by an administrative authority, a patent can still be challenged by third parties. They will have to do so in front of courts, requesting the patent to be *revoked*. The patent holder also has to go to courts in order to enforce his or her patents, alleging *infringement* by third parties.

Patents are only one instrument for the protection of new creations. Other types of intellectual property rights are available, with their own specificities:

- Utility models are a sort of 'petty patents': They have similar characteristics to patents (protect technical inventions, provide exclusivity for a limited period of time; etc.), but to a lesser extent (shorter duration, etc.)

- Design patents protect the aesthetic aspects of goods. They protect new, original, and ornamental designs for articles of manufacture; they are generally less valuable than utility patents because they only protect the appearance of a product and are easier to design around.

- Plant variety protection: It allows the buyer (of seeds) to use second generation seeds for himself or herself and to use these seeds for further improvement without any licence to the initial breeder. But he or she cannot sell the second generation seeds to third parties.

- Ad hoc rights have also been set up, such as the Chip Protection Act (CPA) of the US (1985).

- Copyright protects expression of ideas (original works of authorship), while patents protect ideas or inventions. Hence copyright is narrower than patents. There are many exemptions to copyright ('fair use', etc.), more than to patents. The duration of copyright is longer than patents (up to 70 years after the death of the creator). Certain borderline cases can be protected by both copyright and patents (e.g., software and genetic code).

- Trademarks protect names, words, symbols, or devices. The purpose is to ensure proper information of customers and encourage investment in quality of goods and services. Trademarks inform the customer about the origin of the products and distinguish them from other goods. Trademarks have no time limit, but must be maintained by their holder.

- Trade secrets may consist of any formula, physical device, or compilation of information that provides its owner with a competitive advantage and is treated to prevent others from learning about it (Coca-Cola formula, source code, recipes, customer lists, etc.) Trade secrets have no time limit.

- Trade dress is a distinctive (non-functional) feature which distinguishes a merchant's goods or services from those of another (colours, textures, shapes, packaging, etc.)

BASIC ECONOMICS

Economists consider patents to be an incentive mechanism, aimed at eliciting investment in inventions and innovations, and in the effective use of these inventions. Patents give to their holder a right to exclude third parties from using the invention. On the basis of this restricted competition, the patent holder can exercise some power on the market and charge customers a mark up on the price. In the absence of this exclusion right, competitors could imitate the invention while they have not incurred the research cost, and therefore offer the good at a lower price than the inventor. As inventors would then be driven out of the market this situation could deter inventions coming to the market place in the first place. The down side of patents is that they impose on customers a price which is higher than the marginal cost of production, hence generating an economic inefficiency, a 'deadweight loss'. Customers ready to pay more than the marginal cost but less than the mark up price cannot buy the good, although they would compensate society for the resources used in producing that good (i.e. its marginal cost). This produces the so-called 'Schumpeterian dilemma' between static and dynamic efficiency. The exclusion right creates some kind of monopoly which is a static inefficiency, while fostering innovation which is dynamically efficient.

Patents are a double edged sword with regard to the diffusion of technology. On the one hand patent documents must include a clear description of the invention, hence patents disclose inventions which could otherwise be kept secret. On the other hand, patents erect barriers to the use of knowledge, which becomes more expensive as third parties have to pay fees to use the patented invention. The disclosure role of patents is irrelevant in the case of inventions which should have been publicized in any case (e.g., new products which are provided to customers as opposed to new processes which can be kept secret). It is the purpose of patent policy to design a system that maximizes the benefits and minimizes the costs for society.

1.3. **Recent Trends: Patents in the Knowledge-Based Economy**

The number of patent applications to the EPO has increased by 150 percent between 1995 and 2006, passing from 80,000 to 200,000. That makes a yearly growth rate of more than 9 percent, compared with less than 2.5 percent for GDP in Europe during the same period of time. This figure of 9 percent is somewhat exaggerated as it includes PCT applications, of which an increasing part is never transferred to the national or regional stage, and never becomes a real patent application. If one excludes non-transferred PCT applications the figure is still very high, however. The number of applications to the EPO increased from about 60,000 a year in 1990–94 (a period of stagnation) to about 110,000 in 2000–02 ('priority year', i.e., the year an application is first filed) meaning a growth of more than 80 percent. This makes an average annual growth rate of nearly 7 percent a year, similar to the growth rate of applications and grants at the USPTO. These figures are all the more impressive when put in a historical perspective. Looking at the US, the only country for which such a long data series is available, one has to go back to the 1870s to find such a high growth in patent numbers.

SO WHAT HAPPENED?

Business R&D expenditure among OECD countries grew by about 50 percent between the early 1990s and early 2000s, while at the same time GDP increased by about 25 percent. This is often overlooked because total R&D has not increased that much in the OECD, but business R&D has increased very rapidly, while government R&D in that period has shrunk. Business R&D has surged notably in particular areas, such as biotechnology and information technology, which also contributed 50 percent of the increase in patenting over that period. However, the surge in business R&D, which was 50 percent, does not explain entirely the 80 percent increase in patent applications. Overall, the 'propensity to patent' (the number of patents taken per dollar or euro of R&D, assuming the productivity of R&D constant) has increased by 20 percent in less than 20 years in the OECD (following a long-term decline in the US). This increase is due to technological change, to economic transformations; and to patent policy shifts that took place during this period (OECD 2004b).

The emergence of the Internet and all means of communication have boosted the extent and speed of the circulation of knowledge. It is probably more difficult to keep an invention secret nowadays than it was two decades ago. As they cannot rely as much as before on secrecy, inventors have to use legal means of protection instead.

In many countries important sectors of the economy have been deregulated resulting in more intense competition. Also, due to globalization, national markets are opening to foreign entrants, hence becoming more competitive. Competition generates demand for patents. National monopolies do not need patents, as nobody can imitate them, whereas companies exposed to competition are also exposed to imitation and need patent protection. In the mean time there have been changes in the way innovation is organized. The dominant model used to be large companies innovating in-house. Now, under pressure from competition and financial markets, companies are increasingly specialized. They don't have all the technology and competences they need in-house; they have to obtain it from others, through purchase and alliances. Such contractual relationships take inventions out of the company and put them on the market place. It requires all parties to protect their intellectual property to delineate clearly what they want to share or not and under which conditions. A related trend is the growing importance of small, high tech start-up companies. They do not have control of, or even access to, distribution channels, they do not have brand names, and they do not have large manufacturing facilities. They have one thing: their technology, and they have, in general, no other way of protecting this technology than by filing patents.

At the same time, the use of patents has broadened and diversified. Licensing is becoming more widespread; cross-licensing has invaded entire industries, like semi-conductors and many other fields of electronics and, more recently, biotechnology. Patents are used to raise capital. They are a quality signal of the technical advance to capital markets or to venture capitalists; they serve as collateral for bank loans; or they are even securitized. Such new uses have made patents more attractive. In a nutshell, the economic transformations of the last decades have all increased the demand for patents.

A further factor causing a surge in patent numbers is the 'pro-patent' policy stance followed in OECD countries over the last 25 years or so. This stance started in the US in the early 1980s when the Carter and Reagan administration, frightened by the progress made by Japanese companies, decided on policy packages aimed at regaining the US technological dominance. One set of measures was to strengthen the patent regime in the US. As the US was apparently successful in strengthening its technological position between the early 1980s and the mid-1990s, other developed economies followed a similar pathway. At the same time, keen to protect themselves from imitations made in developing countries, be it through counterfeiting or production of generic drugs, the US and other developed countries exercized pressure for a worldwide upward harmonization of intellectual property protection standards. That resulted in the TRIPS (trade-related intellectual property rights) agreement being signed in 1994, which includes a set of minimal standards that all signatory countries should set in place and which is enforced by the World trade Organisation (WTO).

The first measure taken in the US was to strengthen the governance of the patent system and this was followed by other countries. A central court of appeal (CAFC) was created in 1982 which would influence both standards and enforcement of patents in the US for the next decade—generally in the direction of making it easier to obtain patents and to enforce them. Japan set up a central intellectual property (IP) court in 2005 to unify the treatment of patents and litigation. In its own way Europe has also strengthened its patent institutions, by creating the European Patent Office in 1978, and by passing a few European Commission regulations. National laws are still important, and have evolved in a similar upward direction.

Enforcement has been strengthened, with increased damages attributed to patent holders in case of infringement (more than several hundred million US dollars in a significant number of cases). This increased legal value has led companies to file more patents.

The patentable subject matter has been expanded to include progressively genetic material; software related inventions; and (in the US) business methods. This trend has been led mainly by court decisions, sometimes later converted to acts by legislative bodies.

Exemptions to patent enforcement, related notably to research use, have been narrowed down in the US, while they are being questioned in other countries. It was essentially a privilege of universities that they would not be subject to court actions if they used inventions patented by others for research purposes. They cannot do it so easily now. On the other hand, universities have been encouraged to file patents on their own inventions, following the Bayh-Dole Act passed in the US in 1980 and followed in the 1990s by almost all OECD countries.

As a result of these economic and policy changes, it seems that a larger proportion of inventions are patented nowadays than ever before. Patents have become the currency of the knowledge economy.

1.4. **Structure of the Book**

This book is structured around a series of key issues.

Chapter 2 (by Dominique Guellec) addresses the genesis and evolution of the patent system, starting in Venice in 1474 and leading up to the recent international changes by the World trade Organisation (WTO) and the EPO. This chapter attempts also to analyse the emergence of major features of modern patent systems, such as the inventive step or a 20 years duration.

Chapter 3 (also by Dominique Guellec) investigates the rationale of patents and their economic role. It contrasts the 'natural rights' theory with the utilitarian approach on which economics relies. It compares the patent system with other policy instruments aimed at encouraging innovation, such as public laboratories or subsidies.

Chapter 4 (by Dominique Guellec, Bruno van Pottelsberghe, and Nicolas van Zeebroeck) explores the economic use of patents, including gaining freedom to operate or blocking competitors. It looks at the growing market for licensing and analyses the challenges raised by patents to competition policy. It reviews methods for valuing patents.

Chapter 5 (by Dominique Guellec) presents 'patent design', the discipline which studies the substantive features of the patent system (subject matter, inventive step, scope, and duration) from the point of view of their economic effects. It focuses on the European law and practice.

Chapter 6 (by Niels Stevnsborg and Bruno van Pottelsberghe) presents patent procedures at EPO and the filing strategies adopted by firms. It reviews the rules, from application to opposition; how they can be played strategically by applicants; and how the EPO examiners can respond.

Chapter 7 (by Bruno van Pottelsberghe) reviews a series of challenges currently faced by the patent system in Europe: patenting by universities; the cost of patenting; the relative operations performance of the EPO; the potential benefits of a stronger integration of the European market for technology; and the sharp increase in the EPO workload.

Chapter 8 (by Dominique Guellec and Bruno van Pottelsberghe) concludes by drawing a series of policy implications for the European patent system suggested by the economic perspective developed in the previous chapters.

Part 1
The Economics of Patents

2 **Historical Insights**

Dominique Guellec

This chapter gives an overview of the emergence and evolution of patent systems across history, from the first patent law in Venice (1474) to the current situation. It shows how major features of modern patent systems progressively emerged and took the shape they have now; it identifies the most influential determinants of this evolution; and it recalls the public controversies which came with these developments. The first three sections of the chapter adopt a chronological approach, from the emergence of the patent system in the fifteenth century to its latest developments: the TRIPs agreement. The fourth section reports the evolution of major features of patent systems, such as the duration of protection and the inventive step. The fifth section draws lessons from history to modern patent systems.

2.1. **Genesis of the Patent System: Fifteenth to Eighteenth Century**

The emergence of patents is part of the march of markets and capitalism. In antiquity and the middle ages, there was no formal protection of inventions, and 'organized secrecy' prevailed. Certain social institutions, notably guilds, monitored competition and ensured some sharing of know-how and technology among their members, while denying non-members access to that pool of knowledge. Even the transmission of knowledge first took family channels. Such an environment would in general deter innovation, which was seen as a source of destruction more than creation, and which would destabilize established positions, harming incumbents. In the middle ages kings, bishops, and other political powers started employing engineers and highly skilled craftsmen to conduct certain technology-intensive work, such as manufacturing military equipment and building cathedrals. Such works required innovations. Inventors were rewarded by further contracts or a promotion in the administrative hierarchy, and they kept their inventions secret when they could.

Medieval institutions were designed to protect secrecy and, for many of them, to deter innovation. The aim of patents is exactly the opposite, and it is in this hostile context that the notion of patent protection emerged. In fact, patents have an ambiguous relationship with the medieval institutional setting. On the one hand, they are part of the competitive process and are at odds with medieval institutions; on the other hand, they represent an exception to competition: a kind of privilege obeying a similar logic as other privileges that medieval powers were used to grant.

Awareness of a link between some kind of monopoly and a higher reward, and the idea that this link could be used to encourage innovation, arose quite early. In the Greek city of Sybaris in Sicily, 500BC, the king would grant a one year exclusivity to chefs who had invented a new recipe.

Privileges of all sorts were a usual way in the middle ages for the lords and kings to encourage or facilitate various types of industrial or commercial activities. Such privileges could consist in exemptions to taxes and other duties, or in exclusive rights of exploitation. The latter were used notably for underground exploration and mining activities. These concessions involved an exclusive right to explore and exploit the underground, granted for a limited period of time, but no ownership on the land itself. Such mining concessions could be seen as precursors to modern patents (Braunstein 1984).

First privileges rewarding inventions were granted in independent cities of northern Italy in the late middle ages. They were explicitly targeted at attracting foreign craftsmen (in other words, those established in other cities)—those who accepted to settle in the city to perform their art and to train local workers would benefit from five to 20 years of exclusivity. Examples of early privileges include the famous Brunelleschi patent for a system that would transport marble on the Arno river, granted by the city of Florence in 1426. Several such rights were also granted in Venice. This early patent practice was formalized in the very first patent law, the patent statute of Venice issued in 1474.

Box 2.1. The patent statute of Venice (adopted 19 March 1474)

There are in this city, and also there come temporarily by reason of its greatness and goodness, men from different places and most clever minds, capable of devising and inventing all manner of ingenious contrivances. And should it be provided, that the work and contrivances invented by them, others having seen them could not make them and take their honour, men of such kind would exert their minds, invent and make things which would be of no small utility and benefit to our state. Therefore, decision will be passed that, by authority of this council, each person who will make in this city any new ingenious contrivance, not made heretofore in our dominion, as soon as it is reduced to perfection, so that it can be used and exercised, shall give notice of the same to the office of our Provisioners of Common. It being forbidden to any other in any territory and place of ours to make any other contrivance in the form and resemblance thereof, without the consent and license of the author up to ten years. And, however, should anybody make it, the aforesaid author and inventor will have the liberty to cite him before any office of this city, by which office the aforesaid who shall infringe be forced to pay him the sum of one hundred ducates and the contrivance immediately destroyed. Being then in liberty of our Government at his will to take and use in his need any of the said contrivances and instruments, with this condition, however, that no others than the authors shall exercise them.

(Mgbeoji 2003, trans. Luigi Sordelli)

Important features of later patent systems are present already in the Venice statute, some of which still characterize current systems:

- A patent is defined as a right to exclude.
- The aim of the system is to incentivize invention and import of new techniques.
- Patented techniques should be new to Venice ('our dominion'), not necessarily to the entire world.
- The invention should be practical ('reduced to perfection'): no patents shall be given on mere ideas or scientific discoveries.
- There is examination of the patent application (by the 'office of our General Welfare Board'). The usefulness of the invention to the Venetian economy is the primary criterion for assessing its patentability.
- The duration is standardized (to ten years).
- In case of alleged infringement, the patentee will go to court.

The Statute seems to have been rather popular with inventors, as the number of patents granted in Venice increased over time: 33 in 1474–1500, 116 in 1501–50, 423 in 1551–1600. The statute was in force until 1650.

The Venice statute is modern in two ways at least. First, it is a statute, which as such gives guarantees to inventors and their competitors by limiting the discretion of the ruler. This is an essential feature for patents becoming a market instrument. That makes the perspective of getting a patent more predictable ex ante (at the stage of research) by the inventor while protecting existing trade from arbitrary restrictions decided by the ruler. Second, although the import of technology is mentioned, local inventors also are explicitly included into the system. Hence the purpose is not only to capture existing techniques, but to create new ones domestically.

Venice was a city largely controlled by merchants and industrialists, whereas major states in Europe were controlled by royal families and the nobles. When these states progressively set up patent systems, they took inspiration from the Venice statute but the initial stance was less market friendly and more oriented towards the immediate benefits of the state itself, which was mercantilist in spirit. Mercantilist types of practice and legislation were taken in most advanced European states in the mid-sixteenth century: France, England, Germany, Holland, Spain, and Denmark (Prager 1944; David 1992; Belfanti 2004).

In England, notwithstanding early and rare cases back in the fourteenth century, the first concession to an inventor was granted in 1561. The criteria for granting a patent were that the invention would substitute to imported goods and that it would make the good available for a lesser price than the imported one. It should also not substitute domestically produced goods. Hence the economic impact was the primary criterion for granting patents, that the invention should not hurt domestic industry but rather provide it with new opportunities. Patents were granted by the Crown (Privy Council). However by

---

Box 2.2. The British statute of monopolies of 1623: extracts

All monapolies and all comissions graunts licences charters and letters patents heretofore made or graunted, or hereafter to be made or graunted to any person or persons bodies politique or corporate whatsoever of or for the sole buyinge sellinge makinge workinge or usinge of any thinge within this realme or the dominion of Wales, or of any other monopolies, or of power libertie or facultie to dispence with any others . . . are altogether contrary to the lawes of this realme, and so are and shalbe utterlie void and of none effecte, and in noe wise to be putt in ure or execution.

Provided alsoe and be it declared and enacted, That any Declaracion before mencioned shall not extend to any tres Patente and Graunte for the tearme of fowerteene yeares or under, hereafter to be made of the sole working or makinge of any manner of new Manufactures within this Realme, to the true and first Inventor and Inventors of such Manufactures, which others at the tyme of makinge such tres Patente and Graunte shall not use, soe as alsoe they be not contrary to the Laws nor mischievous to the State, by raisinge prices of Comodities at home, or hurt of Trade, or generallie inconvenient; the said fourteene yeares to be from the date of the first tres Patente or Grant of such priviledge hereafter to be made.

(Section 6)

---

the end of the sixteenth century increasingly dubious patents were granted (e.g., one on playing cards), some harming established industry hence infringing on the 'liberty of commerce', while significant inventions were denied protection. The Queen in fact used the system for rewarding her servants while not raising taxes directly, hence she used patents as indirect and privatized tax raising instruments. The clientelist and excessive use of the system by the Crown of England led to discontent, which resulted in regulation of this practice: the British statute of monopolies adopted by Parliament and House of Lords in 1623.

The Statute of Monopolies codifies the grant of patents in line with the Venice statute:

- A patent is an exclusive right and should be considered as an exception to competition—which remains the rule.
- The patent must be granted to the first and true inventor, a broad category which in fact includes importers as later cases would show.
- The invention must not result in the raising of prices and it must not hurt trade.
- The patent is granted for a limited duration, maximum 14 years (corresponding in fact to two terms for an apprenticeship in a British craftsman).

The patent system as designed in the statute has an ambiguous connection with competition. The introduction of new industries in the kingdom, which is the aim, should not come at the expense of established interest. In a way, patents are an attempt by the state to introduce more competition, as they

Table 2.1. Number of patents granted in England

England	1561–80	1581–1603	1617–29	1630–39
Total grants	35	20	33	75
To foreigners	19	2		

Sources: Beltran (2001: 46); Mossoff (2001: 5, 7).

allow circumventing the guilds system, which blocks innovation, especially when promoted by outsiders. Innovations will be permitted, even protected, outside the Guilds. Establishing new industries in the kingdom also means creating new markets, which after a while (14 years maximum) will be opened to competition. The statute of monopolies had apparently a positive effect on patent filings, which increased significantly after its adoption (see Table 2.1).

The Statute of Monopolies was pushed forward by the legislative branch in order to limit the discretionary power of the executive branch. In fact, after the statute was passed patents were still granted by the Privy Council of the King. The Commons fought back for more than one century after the adoption of the statute before they could actually have the patent jurisdiction passed to common law courts which could revoke patents granted by the Crown. As a result, the Crown progressively lost control of the system, which then passed to courts. On that basis, a legal doctrine started evolving, which put emphasis on the interest of the patentee and of third parties, instead of the state, which defined what the rights and obligations of the patent holder are in a competitive context. For instance a detailed description of the invention was progressively requested in England by courts (starting in the 1730s), which should be published at the time of grant. The purpose was to make it clear in case of litigation what was protected and what was not, and to allow third parties to practice the invention when the patent is over. The patent is no longer an instrument in the hands of the Crown for developing the economy; it is a reward legitimately foreseen by a competitive inventor. This evolution took place mainly during the eighteenth century, and was clearly one component of the progress of market institutions in England.

In France, the first reported privilege for a technique 'new to the kingdom' was granted in 1551, as an Italian glass maker was given a ten years monopoly by the King, Henri II, for manufacturing 'Venetian type of glass' (reduced to five years by the Paris Parliament). 'Patentes royales' were instituted in the mid-sixteenth century, with the explicit aim of attracting craftsmen from northern Italy, notably those who specialized in luxury goods. A description of the invention was publicized when the patent expired so that the invention be publicly available. These privileges were similar to later patents: novelty requirement; industrial applicability; average duration of 20 years; punishment of counterfeiters; possibility of granting licensing contracts; etc. (Flynn 2006). National patents were applied to the Conseil du Roi (Royal Council). The Council requested the opinion of experts before granting or refusing the patent. That decision could be opposed by third parties to the Paris Parliament. Experts were initially merchants and craftsmen, gathered in a commission set up in 1602, which was replaced in that role by the Académie des Sciences in 1699. The 'Bureau du Commerce' was charged after 1730 to do preliminary examination of applications before they were submitted to the Royal Council. The final decision was taken by the Contrôleur général des finances, who informed the Royal Council. In addition to national patents, there were local privileges, valid for certain regions only, and granted by the local authorities (Hilaire-Perez 2000).

In the eighteenth century the French patent system was rather different from the English one. In France, patents were an instrument of technology policy in the hands of governmental and local authorities. Patents were used in parallel—as complements or substitutes—with financial grants, subsidies, tax exemptions, and other types of support (e.g., access to public procurement; the possibility of using the cheap labour of hospitals; etc.) The pre-grant examination procedure was more structured in France than in England, but there was little ex-post control by Courts. The decision to grant or not, and the extent of protection (in terms of duration, geographical coverage, and scope) were fixed in accordance with the estimated economic value on an ad hoc basis. Patents could be revoked by the State itself, or simply superseded by the later issuance of other patents to competitors having invented superior technology.

The patent system was born under mercantilist auspices. The role of patents was not to support competitive markets, but to strengthen the economic and technological foundation of the State. However, in order to reach this objective, the Mercantilist State had to rely on a market-type of incentive. Attracting foreign craftsmen and supplementing existing mechanisms (notably guilds) for encouraging domestic innovation were the leverages for increasing the national stock of technology. A more market oriented patent system, based on stable and transparent rules enforced by neutral Courts, was possible only when market institutions (the law and the state) had reached a certain degree of maturity and the volume of market transactions was high enough for sales to compensate the innovator. That happened in England in the course of the eighteenth century, and in France at the end of that century.

The patent system went through a series of advances and crises during the eighteenth century, both in England and in France. Following the abuse of the system for speculative purposes during the South Sea Bubble, it entered into a credibility crisis in the early eighteenth century. Patents were used by their holders to ransom established business and to deceive the public. As a reaction, the requirements for upholding patents were made more rigorous by Courts in England. In France also, as the number of patent filings had started to decline

Table 2.2. Applications and patents in France

	1730–39	1740–49	1750–59
Applications	52	50	25
Grants	30	20	15
Grant rate (%)	60	40	60

Source: Hilaire-Perez (2000: 118, note 114).

in the mid-eighteenth century (see Table 2.2), decisions were taken in 1762 to strengthen the requirements for granting patents and to improve disclosure. With the advent of the Industrial Revolution and the related surge in inventions, the English authorities decided to reduce the cost of patenting and simplify procedures, an example followed by the French in 1776 (who also abolished the preliminary examination). This is an early example of 'pro-patent policy' associated with a wave of new technology. The emblematic invention of the time, the steam engine of James Watt, was protected by a patent which was renewed for 28 years.

2.2. **The Crystallization of Modern Patent Systems: The Nineteenth Century**

The late eighteenth century–early nineteenth century marked a step in the evolution of patent systems. Laws were passed in several countries (France, the US, and then many countries in continental Europe) and administrative procedures were modernized. This was related to the spread of the Industrial Revolution, which gave more importance than before to technical change. It was also related to the emergence of the modern state and the rule of law, and to the strengthened role of markets in the economy. Patents, as other types of privileges, were reformatted in the new institutional framework that prevailed after the French Revolution.

France passed a patent law in 1791, which was revised later in 1844 (Beltran et al. 2001). The 1791 law presents intellectual property as a natural right of the individual, of the same kind as any other type of property. With this law, France abandoned officially the prior review system. The view was that government was not entitled to judge the value of an invention, which should be left to society and done through the market test. If a problem were to occur, as an infraction to competition law, that should be ruled by the courts. The duration was five, ten, or 15 years, depending on payment by the patent holder. The law includes a two years working requirement. Applying for protection abroad would make the patent invalid in France, although inventions already protected abroad could be protected also in France. Modifications made in 1844 include notably dropping the 'not patenting abroad' requirement. The French model was extremely influential in the nineteenth century, before recessing in the twentieth century: practically all European and most of the Latin American patent laws issued in the nineteenth century were modelled on the French law.

In the United States a patent system was established by the patent act of 1790, early after independence, with a basis in the Constitution (1787), which famously states that: 'The Congress shall have Power . . . to promote the Progress of Science and useful Arts, by securing for limited Times to Authors and Inventors the exclusive Right to their respective Writings and Discoveries.' (Constitution of the US, Article 1, section 8, clause 8). That followed existing laws in some of the 13 founding states (e.g., Maine, Massachusetts). The patent system was adopted under strong pressure of industrialist lobbies, claiming that this was needed for strengthening the technological base of the US. The initially established US system included an examination process of patent applications, such that non-novel ones would be excluded. In 1793, it turned out that the three commissioners in charge were overwhelmed by the number of applications they had to process (only 55 patents were granted between 1790 and 1793), so that examination was abolished and a registration system was installed. It lasted until 1836 when, in view of a large number of 'frivolous patents' used by their holders mainly for litigating manufacturers, the US Patents and Trademarks Office (USPTO) was established, with the mission of selecting those applications which would actually deserve protection, through prior examination. The 1790 law also requires publication of the invention in a way that makes it understandable by 'the skilled workman'. The patent lasted for 14 years, and then was expanded to 17 years after the grant, in 1861. It could go only to the 'first and true inventor' (not to a mere importer). There was a period of grace of one year (the invention could have been used 'for experimental purposes' up to one year before application). The US system is, along with the German system, a major source of current patent systems.

In Spain a patent law was passed in 1811 close to the French model. It also included 'patents of introduction': exclusive rights granted to others than inventors for producing in Spain goods patented in other countries. These patents were granted for five years only and were expensive. They would not prohibit the import of the good by others.

The different German states adopted patent laws prior to unification (first was Prussia in 1815; Khan 2006). A single patent law for the entire Zollverein entered into force in 1842. Certain Landers still opted out of the system (Hamburg, Bremen). A more thorough law was adopted in 1877 for the Reich. This law included many new features which then diffused in other countries and still prevail today. The 'first to file' rule replaced the 'first to invent'. Patents were published before grant. Third parties could oppose the patent prior to grant.

There is a preliminary examination, initially conducted by commissioned experts, then (1891) handled directly by staff of the newly created German Patent Office (Deutsche Patentamt, DPMA). The key, new notion for examination was the 'inventive height' (*erfindungshöhe*), the ancestor of the inventive step. Once granted, the maintenance of the patent was subject to the payment of renewal fees. The law also included restrictive clauses, which were quite common at the time, for instance compulsory licences: no patentee could refuse granting licences under reasonable conditions; otherwise it would be deemed anti-competitive behaviour. There is a working requirement in the three years following the grant. Strict enforcement of these rules allowed the DPMA to become the reference in Europe for the quality of its patents. In addition, a high fee level made the German system extremely selective. In order to balance the system, i.e., to give some sort of protection to smaller inventions, Germany invented utility models. These are weak and cheap patents: granted with no examination; with a lower inventiveness requirement; and for a short period of time (a maximum of ten years). Many countries would also adopt utility models later.

Box 2.3. The patent controversy of the nineteenth century (see Beltran et al. 2001)

The 'Patent controversy' developed in Europe in the years 1840–75. It was a time of firmer establishment of patent laws and institutions in many countries, and a number of interest groups which suffered from that, and of economists reflecting what they saw as an undue restriction on competition, organized resistance against this expansion.

Abolitionists challenged the negative effect of patents on competition ('Un outrage à la liberté et l'industrie', in the words of Michel Chevallier, a French economist, in 1850), resulting in less economic efficiency and in a violation of the freedom of individuals to do business. They denied the positive effect of patents on innovation (there are other ways for the innovator to get a reward) and on disclosure. They also attacked patents on moral grounds. How could one attribute a reward to an individual, while the inventive process is collective and ideas pertain to all humanity?

The example of Daguerre was used in the French debate: the inventor of Daguerreotype had been granted a state pension in France in 1839, in exchange for placing his invention in the public domain. The following years experienced a wave of rapid improvement of the initial invention, as at the same time prices were falling—in France notably. The opposite situation prevailed in England, where Daguerre (and his local competitor Talbot) had patents—which apparently blocked the progress of photography there for more than a decade.

The debate resulted in the abolition of the Dutch patent system in 1869, and in delaying the adoption of patent systems in the German Zollverein (with the support of Bismarck) and Switzerland. Heated debates took place at the UK Parliament and in France.

The controversy faded after 1875, as a result of three driving forces: i) the patent system was continuing its expansion in most countries; ii) large numbers of inventions related to the industrial revolution of the late nineteenth century were filed to patent offices (Thomas Edison filed more than 1000 patents at the USPTO); and iii) international pressure for harmonization of patent systems was mounting. Pro-patent industry lobbies succeeded in saving the patent system.

The Dutch patent law was voted in 1817. It was revoked in 1869, following the patent controversy (recognizing the anti-competitive aspect of patents, which was a shock to a people of merchants, see Box 2.3) and strong lobbying by the Dutch industry, with a strong component of traders. The patent system was reinstated in 1911: but it remained extremely restrictive; it was actually the most selective in the world with a grant rate close to 10 percent, reflecting the reluctance of the Dutch authorities vis-à-vis patents.

Switzerland passed its first patent law in 1888. It covered only inventions which could be represented by a mechanical model, hence notably excluding chemical industries (which were actively imitating their German competitors). A 1907 law extended patents to all technologies (notably under German pressure).

The English system evolved more smoothly during the nineteenth century as it was already quite solid at the beginning of this period. It continued referring to the Statute of Monopolies (as it would until 1977). Case law then was the engine for the system. The law of 1852 established the UK Patent Office with the mission of handling the entire application procedure (this had been split previously between several administrations which made it extremely cumbersome and costly) and to disseminate the knowledge embodied in patents. The 1902 law introduced prior examination for novelty.

In Japan, the establishment of patent law was faced with resistance: a first law was enacted in 1871 (Meiji period) but immediately revoked due to pressure from all national quarters. A second law was passed in 1885, which proved more durable; it included two years' local working requirement. Foreigners could not get patents until the 1899 law (voted under foreign pressure). Initially based on 'first to invent', the system passed to 'first to file' in 1920, when a strict examination process was also established. Patents granted were very narrow (no more than one claim was allowed). This reflects the role of patents in the Japanese economy, which is consistent with the way the economy works in many respects: it is not explicitly a competitive weapon—it is a way to signal to others that you are active in an area, and to secure freedom of operation within small areas of the market (while not preventing others from entering neighbouring segments).

2.3. **Patent Systems Since 1900**

Most advanced countries at the end of the nineteenth century were equipped with substantial and robust patent laws and institutions. Patents had become part of the standard institutional package for a modern economy or one which aspired to modernize. The twentieth century was mainly characterized by international, upward, harmonization of patent law and practice, and by growing cross-country cooperation in patent matters. Within this global trend, after the Second World War Europe has been building its own regional system, with a common legal background (the European Patent Convention) and a centralized examining body (the European Patent Office). The last decades of the twentieth century were marked by a significant strengthening of the rights associated with patents in most parts of the world.

INTERNATIONAL HARMONIZATION

Since the late nineteenth century international harmonization has been the hallmark and a major driving force of patent systems worldwide. In the nineteenth century all countries had provisions in their patent law discriminating against foreigners. That included higher fees (in the US until 1862) or a working requirement on the national territory (patents would not cover imported goods—e.g., in France until 1845). With the expansion of cross-borders economic relationships—flows of goods, capital, and technology—these provisions became increasingly burdensome to business, and industrialists increasingly pushed for international harmonization. Following the international exposition of Vienna in 1873, where US companies raised concerns regarding the protection of their intellectual property in Austria (which had to strengthen its patent law as a consequence), international discussions started, led by France. The first outcome was the Paris Convention of 1883, initially signed by 10 countries (Belgium, France, Guatemala, Italy, Netherlands, Portugal, San Salvador, Serbia, Spain, and Switzerland). It was then joined by the United Kingdom in 1884; the US and Mexico in 1887; and Germany in 1903. The Paris convention establishes the principle of national treatment, whereby foreign residents should be treated in any country the same way as nationals. This was preferred to the principle of reciprocity (foreign residents would be treated in any country as foreigners are treated in their own country) favoured by the US. The Paris convention also established the notion of 'priority application', by which an applicant in any country is potentially protected worldwide for a certain period of time (initially six months, later one year) before choosing to file protection, or not, in other countries: such a notion is a pre-requisite of reliable international protection and for a properly defined prior art.

The Paris convention has been revised six times since its inception, the last time was in Stockholm, in 1967. Its implementation was monitored by the Bureau de l'Union de Paris (or International Union for the Protection of Industrial Property). This body was integrated in the UN system in 1967, giving birth to the World Intellectual Property Organisation (WIPO).

From one revision to the next, the treaty was progressively reinforced. Restrictions to the exercise of patent rights, notably all types of working requirements, were progressively narrowed down. A further drift was taken in the 1980s, when the US started pushing for having patents and IP integrated into the framework of international trade agreements (the GATT). The result was the TRIPs (Trade Related IP) agreements, signed in 1994, which is now monitored by the World Trade Organisation (WTO). This was a major step as the WTO has a power of punishment that WIPO bodies do not have, and hence IP law becomes a much more serious matter. The essential features of the TRIPs are: patents should apply to all fields of technology; the minimal duration for patents is 20 years after filing; and no working requirement should be imposed except in exceptional circumstances. That imposed a major change to many developing countries, which were given 10 to 15 years to adapt their legal framework. The Indian parliament passed an entirely new patent law in 2005.

In view of the increasing numbers of applications abroad, the National Association of Manufacturers (NAM) of the US launched in 1965 the idea of an international patent procedure which would save on cost and administrative burden for applicants. In 1970 the Washington Treaty was signed which instituted the Patent Cooperation Treaty (PCT) which was implemented on 1 June 1978 (the same day as the EPO was officially set up). That treaty instituted an international procedure for patent applications. The applicant must file an application to the WIPO with a list of countries where he or she would be willing to validate the application later. This application is then searched by one of the patent offices labelled by the WIPO, called an International Search Authority (ISA). The list of bodies forming the ISA includes notably EPO, USPTO, and JPO). However, examination, and granting of a patent, remain under the absolute responsibility of national or regional offices in their respective jurisdictions. That procedure allows applicants the benefit, in addition to the priority year of the Paris convention, of an additional time of up to 18 months during which they choose whether or not to file a patent in any particular country. In addition, applicants benefit from the search report, which gives them information on the likelihood of their having the application granted; which claims are more or less robust; etc. The PCT has experienced several changes since its inception, all making it friendlier to the applicant and ever closer to being a real patent application. On 1 January 2004 the search report was complemented by a preliminary, non-binding opinion of the examiner regarding the prospect for patentability of the invention, called the WOISA (written opinion of the ISA). The WOISA already includes certain aspects of examination. Certain countries are actually supporting the view that the PCT should become a real patent granted with worldwide validity.

In addition, negotiations have been conducted at the WIPO regarding the harmonization of national administrative procedures, resulting in the Patent Law Treaty (PLT) of 2004. Further negotiations started in 2004 in view of harmonizing substantive patent law, in other words, core notions such as the patentable subject matter, inventive step, the disclosure requirement, etc. Convergence between developed countries was achieved, reflecting the similarity of law and practice among these countries (with exceptions such as the patentability of business methods, allowed in the US but not elsewhere). However the negotiations were blocked as certain developing countries led by Brazil would request more stringent disclosure for certain matters (the origin of biological material used for genetic inventions) and protection of 'traditional knowledge', both strongly resisted by the pharmaceutical industry and the US.

EUROPE

The idea of European integration in the field of patents has followed similar path as the whole European project. The Council of Europe, back in 1949, advocated the creation of a European Patent Office. The diversity of patent regimes in Europe was seen as an obstacle to the desired market integration which led to the Treaty of Rome. By 1959 discussions started between members of the EC. The Strasbourg Convention on the Unification of Certain Points of Substantive Law on Patents for Invention was signed in 1963, as a milestone in the process. By that time, in fact, two separate processes were running: a centralized granting procedure for EU members and non-members, and a unified patent system for EU members only.

The former process resulted in the European Patent Convention (EPC), signed in 1973 in Munich and which entered into force in 1977 with 16 countries. The European Patent Office (EPO) was established, on the basis of the EPC, as the executive body of the European Patent Organisation—the institution set up by signatory states to implement the EPC. The first patent application was filed to the EPO on 1 June 1978. The EPC merely provides a unitary application and examination procedure resulting in the grant of a bun-dle of national patents valid in countries selected by the patentee. These patents are notably subject to translation in the language of each country and they are subject to national courts, which have sometimes diverging interpretations of the EPC. The Munich convention was negotiated in parallel with the PCT, as countries wanted to ensure consistency between the two, and to make clear that the EPO would be a major pillar of the emerging worldwide system.

When it was set up, the EPO integrated the IIB (Institut International des Brevets), a body created in 1947 by France and Benelux countries, with the role of conducting novelty search (no examination) for applications filed in these countries. The EPO is a stand-alone body, unrelated to other European institutions such as the European Commission. The European Patent Organisation is governed by an administrative council made of representatives of its member states. The EPO is self-financed, its funding coming partly from fees paid by applicants and partly by 'payments made by contracting states in respect of renewal fees for European patents levied in these states' (EPC: article 37), at a

Table 2.3. Members of the European Patent Organisation (EPO)

Country	Date Joined EPO
Austria	1 May 1979
Belgium	7 October 1977
Bulgaria	1 July 2002
Switzerland	7 October 1977
Cyprus	1 April 1998
Czech Republic	1 July 2002
Germany	7 October 1977
Denmark	1 January 1990
Estonia	1 July 2002
Spain	1 October 1986
Finland	1 March 1996
France	7 October 1977
United Kingdom	7 October 1977
Greece	1 October 1986
Hungary	1 January 2003
Ireland	1 August 1992
Iceland	1 November 2004
Italia	1 December 1978
Latvia	1 July 2005
Liechtenstein	1 April 1980
Lithuania	1 December 2004
Luxemburg	7 October 1977
Monaco	1 December 1991
Netherlands	7 October 1977
Poland	1 March 2004
Portugal	1 January 1992
Romania	1 March 2003
Sweden	1 May 1978
Slovenia	1 December 2002
Slovakia	1 July 2002
Turkey	1 November 2000

current rate of 50 percent. The EPO budget should be balanced: 'The amounts of the fees and the proportion of repayment on renewal fees) shall be fixed at such a level as to ensure that the revenue in respect thereof is sufficient for the budget of the Organisation to be balanced' (Art. 40). The EPO had more than 6000 employees in 2006 (and less than 1000 when it was set up in 1978), of which about 4000 are examiners.

As an international treaty, the EPC is extremely difficult to revise because it requires the unanimity of signatory states. The EPC was slightly revised in 1993, and more broadly in 2000. Membership to the EPO increased up to more than 30 states in the mid-2000s.

The second route to European integration is the Community Patent Convention which would be the basis of a truly European patent. Efforts in that direction were started in the early 1960s, but have repeatedly failed. The Luxemburg Convention on the Community Patent was signed in 1975, but was never ratified. An agreement was reached in 1989 among EU member states, but was never ratified. The third and most recent attempt was led by the EC following a green paper on patents published in 1997. After a general agreement was reached between the EU states in 2003, negotiations were derailed again in 2004 due to translation issues: time delays for translation of the claims into the various EU languages and questions on the legal validity of patents in case of translation errors (for example, what language would be the reference for the courts?)

What is missing in order to have a unitary patent system in Europe is a unitary enforcement system and simplified linguistic rules for the national validation of patents, which would reduce the cost of patenting. EPC states have been attempting to establish rules which would result in an overall system close to that of the Community Patent. The two pieces for that purpose are the European Patent Litigation Agreement (EPLA) and the London Protocol regarding translation. The former would establish a centralized court system in Europe for patent related matters (including revocation and infringement cases), while the latter would simplify translation requirements (i.e., signatory states would not necessarily request that a patent be translated in their own language in order to be validated, but rather a language they have designated from a choice of English, French, or German). None of these agreements was yet in force in 2006.

Overall, the EPC has had a huge effect on the patent system in Europe and is widely seen as a success. First it has strengthened the patent systems in most countries by fostering upward harmonization. Many countries had a weak patent system before the EPC (e.g., no examination and uncertain enforcement), and this is less the case now. The duration of protection was increased to 20 years from application, from a previous 17 years in most countries. Second, the EPC has made it more secure and less costly to obtain protection Europe-wide, by providing standard rules and a centralized, cost-efficient and high quality grant procedure. European integration is, however, clearly not complete, as both separate Court systems and separate language requirements maintain barriers between countries. An imperfectly homogenous interpretation of patent law and the accumulation of national fees and translations reduce the legal certainty and increase the cost of patenting in Europe.

THE REVIVAL OF PATENTS AT THE END OF THE TWENTIETH CENTURY

Patent regimes were strengthened worldwide over the twentieth century. However, the pace of change has been uneven across countries and over time. In the lead country, the US, the role of the patent system has been stabilized, or has even reduced, from the period between the two World Wars to the late 1970s, when it started strengthening again. The strict reading of anti-trust regulation in the US (see the Sherman act of 1890), notably as a follow-up to the 'New Deal' of 1936, led to restricting the rights of patent holders. The share of those patent cases going through the courts which ended up revoked jumped from 33 percent in 1925–29 to more than 60 percent between 1940 and 1954 (Schmookler 1966). In the late 1970s to early 1980s there was a political move in the US which reinstated patents as a central instrument of innovation policy (Kortum and Lerner 1998; Jaffe 2000). Symbolic of that shift is the statement by the US Supreme Court that, 'Everything made by man under the sun is patentable'. In that context a series of important measures were taken (Martinez and Guellec forthcoming). A federal court of appeal specializing in patents was set up—the Court of Appeal of the Federal Circuit (CAFC)—in 1982. This court has been instrumental in reducing the share of patents revoked by courts (from two-thirds to one-third); making patents easier to obtain (by weakening the non-obviousness requirement); broadening the patent subject matter (notably to include business methods); and increasing the damages awarded to holders of infringed patents, etc. The Supreme Court also has contributed to expanding the patent subject matter (to genetic inventions in 1981 and to software in 1993). The US has further managed to export its pro-patent stance, by being the driving force behind the TRIPs and the PCT. The Bayh–Dole Act, in 1980, allowed (and in fact encouraged) government sponsored research entities to take patents on their inventions.

The reasons for strengthening the patent regime in the US are manifold. First was the sudden recognition that the position of the US as the worldwide technology leader was under threat notably due to Japanese competition. The US government felt that companies from other countries were accessing and imitating US technology too easily. A second reason is the wave of technological innovation that emerged in the last two decades of the twentieth century, notably in information technology and biotechnology. That wave fostered the economic value of technology, hence the value of patents, creating incentives for their holders to strengthen the patent system. As a result of these changes, in the 1990s the US experienced its fastest growth in patent grants since the 1870s. The system is more and more widely seen as feeding excesses, such as 'patent trolling' (i.e., attorneys blackmailing operational businesses by threatening them with litigation over dubious patents) or 'patent thickets (i.e., the foreclosure of a market by a company erecting a dense web of patents). As the USPTO became overloaded by exploding numbers of applications, more and more voices would express the willingness to reign in the system, notably through legislation at Congress.

Japan also experienced a radical shift in patent policy during the 1990s. The Japanese patent system used to encourage diffusion of technology and incremental inventions (Ordover 1991). It did so by allowing a long lag between the delivery of the search report and the request for examination (so that applicants could still negotiate (between them) on the scope of their rights); by restricting application to one claim only; and by fostering a system of utility models. However, during the 1990s the Japanese government realized that in order to encourage more radical technological innovations it needed a stronger patent system, and it took much inspiration from the US example. Several measures were taken, such as the possibility of filing multi-claims applications (back in 1988); the reduction of the period for requesting examination (from seven to three years); the creation of a specialized, centralized court of appeal for patent matters (and in fact all IPR matters) in 2005; and the weakening of the utility model system (by suppressing examination, hence weakening their legal validity) in view of encouraging inventors to file (stronger) patents. Japan is also putting much pressure on China to strengthen its patent system.

The emergence of China as a world technology power is a major event of the early twenty-first century. China has been progressively strengthening its patent system since the early 1980s as it has rapidly increased its efforts in research and development, in both public and private sectors. However, as the Chinese industry has been developing its technological capabilities, it has also evolved a capacity to imitate foreign technology, hence infringing on patents, most of them pertaining to foreign companies. The Chinese patent law is of a similar standard as that of developed countries, but its weak point is enforcement: there is a scarcity of judges knowledgeable in patent matters and legal decisions are often not enforced. This reflects to a large extent the state of the court in China in all areas, and it has led to much pressure from abroad—the US, Japan, and the EU—for progress on that side. The Chinese government itself has put much emphasis on strengthening its patent system, not only in order to fulfil its obligations as a member of the WTO (then constrained by the TRIPs), but also as part of a policy stance aimed at promoting innovation by Chinese companies themselves, which increasingly need protection for the outcome of their inventive activities.

Box 2.4. Industrial property in socialist countries

In Russia a patent law was instituted in 1812 and was in place throughout the nineteenth century. Following the revolution of 1917 a new law was passed in 1924 as part of the New Economic Policy (NEP), a transitory liberal turn of the soviet regime. It was modelled on German law, hence quite patent friendly. However, 'authors certificates' were created in a 1931 law, as a complement to patents. This certificate was a kind of diploma and was a reward given to an inventor by the state. The exclusive right to exploit the invention would remain with the state, as there was no free enterprise. The inventor could choose, in theory, between a patent and an author's certificate. In reality, applying for the former was subject to an authorization by the state council, and patents were reserved to foreign companies. In the late 1950s there were about 100 patents and 5000 author's certificates granted per year. Patents make little sense in economies where the role of the market is marginal and exploitation of any technology essentially reserved to the state.

2.4. Archaeology of the Patent System: How its Major Features Emerged

We are examining in this section the progressive structuring of the major features of patent systems: the patent subject matter; restrictions put on authorised owners; duration; fees; procedure (prior examination and opposition); inventive step; disclosure; and limitations regarding the validity and use of patents (i.e., working requirement, compulsory licenses, trading rights). These features emerge only progressively, from new international treaties or from changes in law, in case law, and in practice, and they are still evolving today.

SUBJECT MATTER

The current definition of the patentable subject matter, as defined clearly in the TRIPs and so valid worldwide, includes all fields of technology. Major exclusions are scientific discoveries, mental methods, pure business methods, and artistic creations. Most recent debates involve not the contents of patent law but the categorization of borderline cases, and hence, explicitly or implicitly, the definition of technology.

The only country where patent law has included all technical fields from the outset is the US. The first law to explicitly define what could and could not be patented was the French patent law of 1791, which gave a list of exceptions: scientific discoveries without practical application; and medicines and items covered by copyright. A series of patents covering financial methods were granted in 1791 (147 in total), but this field was excluded in a bill passed in 1792 with the view to discourage financial fraud and speculation, of which the state was a victim (Beltran et al. 2001). Patents on mental methods were granted (e.g., a method for accelerated writing in 1791), but this does not seem to have been persistent. The exclusion of medicines was very common until recently (the German law of 1877 as well as the Japanese law of 1888 excluded them). Only in 1944 did France allow patents on drugs, still restricted to the process and excluding the product itself; in 1960 a special (weak) patent was created for drugs, which were finally made fully patentable in 1968. Several European countries, such as Italy, extended patentability to drugs only in the 1960s and later. This was one of the major changes introduced by the TRIPs to countries such as India, which until 2005 had only process patents on drugs. Chemical products also have had a changing treatment over time. They were excluded from German patent law in 1877 (only processes could be patented) as Germany was trying, successfully, to imitate the then most advanced industry in Britain. The first patent law of Switzerland entirely excluded chemicals, with a view to imitate the German technology. In turn, England, between 1919 and

1949, excluded chemical products from patent protection, with the view to protect itself from the now superior German industry. Japan did the same in 1921. One argument often repeated for banning chemical products from patentability is that they are not inventions, but just discoveries—as these molecules already exist, or could exist, in nature. The German law of 1877 and Japanese law of 1877 also excluded food products (although not the processes for producing them).

The concerns that are behind the limitations in the patent subject matter are of different kind. Some are related the core mission of the patent system—encouraging invention—such as the exclusion of scientific discoveries or pure business methods (it is deemed that their patentability would not encourage their production); others are related to concerns with basic needs of the population and the ability of the nation to be independent in that regard (e.g., food and medicines); finally, some are simply strategic choices by national government in order to access other countries technology at lower cost (i.e. to favour imitation of such things as chemicals). International agreements of the twentieth century have aimed at mitigating the last category, and to a certain extent the second one. These changes have operated under a similar logic as trade agreements—mutual concessions for ensuring mutual (but not necessarily equal) gains.

The fact that, following the TRIPs agreement, the patent subject matter includes all technical fields does not mean that there are no problems any more. With the expansion of science and technology new fields appear, where opposite economic interests confront each other, and whose status in view of patenting criteria is not always immediately clear-cut. Major debates since the early 1980s, both in the US and in Europe, have turned around genetic material and software. It has been alleged that genes are not inventions but only discoveries, hence not patentable; and that software is essentially made up of algorithms and abstract mathematical ideas, hence not patentable. In both cases, the view has prevailed in the courts that where items are of a technical nature then they are patentable, although certain restrictions would apply (especially in Europe regarding software). The issue is different with business methods, which are not new subject matter (there have always been inventions of that type). Their inclusion in the list of patentable objects in the US since 1998 (State Street decision of the CAFC) meant a substantial revision of certain principles on which the patent system was based.

WHO CAN OWN A PATENT?

There have been, across history, various types of restrictions regarding who is entitled to own a patent. These restrictions have been progressively abandoned to converge towards an internationally agreed rule that the legitimate owner of a patent should be the first to file, whatever his or her country of residence, with the exception of the US, where it is the first to invent. Whereas the early patent systems encouraged foreigners to file for protection (as knowledge could then travel only with people), later evolution led to discrimination against foreigners. Techniques would increasingly be published, and their circulation would not require that experts should travel as much. Or, if that was necessary, people having the skills could be attracted with incentives other than patents (e.g., direct monetary compensation) and would serve directly established businesses in the country. The English Statute of Monopolies grants entitlement to the 'first and true inventor', which in fact includes importers (who have not invented the technique, but have introduced it into the country). The French law of 1791 allows foreigners to own patents, although the 1844 law limits the validity of a patent in France until after its lapse in the country of origin. In the US, patents are given to 'the first and true inventor' in the world: the 1793, 1800, and 1832 US statutes restricted patent property to US citizens or residents who intended to become citizens. That would allow free use in the US of inventions patented in other countries by these countries' nationals. In 1836 this provision was removed and replaced by discriminatory fees (significantly more expensive for foreigners than for US nationals). Equal treatment of all was instituted in 1861. The Prussian law also excluded applicants other than 'German Burgers' from patent ownership. The German system set, in the 1877 law, the 'first to file' principle. In Japan, the 1888 law excluded foreigners, but it was amended in 1899 when Japan joined the Paris convention. The progressive abandonment of discriminatory clauses in the course of the nineteenth century reflects the increasingly international nature of technology markets and pressure from leading countries on less advanced ones.

In addition to these restrictions on foreigners, some countries limited the collective ownership of patents. Ownership of patents was limited to five investors (later extended to 12) in England by the Bubble Act of 1720 (following the South Sea Bubble bust), which was upheld until 1832. The motivation was to limit the use of patents in speculative endeavours, which was facilitated when owners where not individual inventors or entrepreneurs. In France the 1791 law, following earlier practice inspired by the British example and a painful national experience with speculation, forbade patent ownership by 'sociétés en nom collectif'. This clause was removed in 1806.

Trading rights on patents (changing ownership) has been also subject to restrictions. In France, under the Ancien Régime, administrative permission had to be obtained first for trading the rights (Hilaire-Perez 2000). Why such restrictions? A patent conveys some monopoly power, which could be used in ways detrimental to society. This is more likely to happen if the holder is not the inventor. Those who purchase existing patent rights could be suspected of expecting to increase the value extracted from such rights in ways unrelated to the value of the technology itself. For instance one could constitute portfolios of patents on substitutable technique which together give disproportionate market power. Also, as illustrated by debates in the US in the mid-2000s, so-called 'patent trolls' (in general, patent attorneys purchasing a patent portfolio and chasing established companies to pay them fees for alleged infringement to their patents) could use low quality patents in clever legal strategies. Restrictions on trading of patents were evoked as one possible solution.

DURATION

The statutory duration of patents is now fixed to a minimum of 20 years from application, in conformity with the TRIPs. It took several centuries before such harmonization emerged. The first step was to move from discretion by the granting authority, which would act on a case by case basis, to national standards. The second step, in the twentieth century, has been to reach international uniformity. In Venice the duration was ten years. In the Statute of Monopolies it was 14 years. In Spain from the sixteenth to the eighteenth century: ten to 40 years. In France there was no standard until the late eighteenth century: the duration of the monopoly was left at the discretion of the granting authority on the basis of the assessed usefulness of the invention to the economy. 'Patentes royales' would vary from five years to perpetuity (or the lifetime of the inventor), and could be extended to one region or the entire kingdom. In 1762 the king of France abolished perpetual privileges and set an upper limit of 15 years, which applied inter alia to patents. The 1791 law fixed the duration to five, ten, or 15 years from application, depending on payment of fees by the assignee (at the time of grant). In Germany the duration was 15 years from filing, in 1877; this was extended to 17 years in 1923. In Massachusetts in the eighteenth century the duration would depend on the alleged value of the invention, being between seven years and the life time of the inventor—with a ten years average. In the US law of 1790, the duration was 14 years, which was increased to 17 years (after grant) in 1861; the Patent Act of 1995, which implemented the TRIPs rulings in the US, moved the duration to 20 years after application. In addition to this standardization of the duration, most countries have implemented systems of renewal fees; in order to maintain the patent in life the holder must pay annual fees. The US adopted the system in the early 1980s, with fees payable at the third, seventh, and eleventh years. That makes the actual lifetime of a patent the choice of the patentee instead of the government, as it was the case initially.

FEES

Fees for filing a patent were initially very high, and that was essentially set deliberately: patents were a way for government to raise taxes, and high fees were also seen as a mechanism for selecting good applications, on the premise that those with low value inventions would be deterred. In Germany in the late nineteenth century, fees were deliberately set high in order to discourage small inventions (or to orient them towards the cheaper utility model protection). Progressively in the nineteenth century, and even more in the twentieth century, lowering fees have become a policy objective for patent offices. This view prevailed in the US from the outset: as the patent system is a way to encourage innovation, it should be made as accessible as possible to all inventors, which notably involves lower fees. This view has become dominant now among patent offices. In addition, some offices (especially in the US and in Japan) offer discount fees to certain applicants who are viewed as financially weak (individuals, universities, and small and medium enterprises). The limit to lower fees is set by the constraint, explicit or implicit in all countries, that patent offices should be financially self-sufficient.

PRIOR EXAMINATION

There are two polar ways of organizing a patent system. One way is to screen ex ante good applications from bad ones, through examination; the other way is to grant all applications through a simple registration system (provided that some formal conditions are met) and leave the sorting to the courts, ex post, if the patent is challenged by third parties. History has oscillated several times between these two models. In the mercantilist system the state would pick inventions it deemed useful to the economy and keep the rest out. This model prevailed in England until the early eighteenth century; and in France until the late eighteenth century. In England a court-based system progressively prevailed, on the ground that the screening by state authorities was often influenced by the particular interest of the Crown and its clientele, then unfair to applicants who complained through the House of Commons. In France prior examination was prevalent, following a first case in 1612 (Hilaire-Perez 2000) under pressure of the Parliament. Examination in France in the eighteenth century was conducted by the 'Bureau du Commerce' (the administration in charge of sponsoring the development of industry), which requested help from the Académie des Sciences regarding technical aspects. However, reluctance to examination progressed at the end of the eighteenth century for similar reasons as those in England, as entrepreneurs were getting stronger and more willing to reduce interference from government, which would use economic and political criteria for assessing applications. The solution brought by the French law of 1791 was to abolish prior examination. By the middle of the nineteenth century, nearly all European states had abolished prior examination, with the exceptions of Sardinia, Belgium, Holland, and Prussia (Beltran et al. 2001).

The US law of 1790 set up a committee of three high level officials, the Secretary of State, the Attorney general and the Secretary of War, in charge of selecting applications. But the committee was immediately overloaded with applications and a registration system was instituted three years later. The 1836 patent law established the US patent Office, whose trained experts would examine applications. This was a reaction to the flow of bad patents, used for legalising imitation or for ransoming legitimate businesses. In the UK, examination was proposed in the 1852 bill but rejected by the House of Commons, on the grounds that it would give too much power to public servants and that there were not enough of them anyway. Only in 1905 did the British introduce substantive examination for novelty, following the German example. Germany in the 1877 law established a requirement for strict prior examination, done by external consultants. But due to conflict of interest (external experts could have connections with the applicant or its competitors), examiners became permanent employees of the Patentamt in 1891. France has kept the registration system up to now, although its close integration into the European patent system has changed its working: Filings to the INPI are searched by the EPO, on behalf of INPI, but not examined regarding the inventive step.

A complementary procedure is opposition, which allows third parties to oppose a patent grant before it is finally issued. It is a kind of low cost litigation procedure, a middle way between examination and real litigation. Opposition is handled by the patent office itself. Germany was the first country to implement opposition, in its 1877 law. The UK introduced such a procedure in 1883 (to be filed by interested parties within two months after the filing of the patent). Japan adopted the system in 1921; until 1996, even pre-grant opposition was possible there (then abandoned due to abuse by certain companies which systematically used that procedure in order to slow down the issuance of patents to their competitors). The post-grant opposition procedure is one of the pillars in the quality policy of the EPO, which adopted it as soon as it was created.

INVENTIVE STEP

In all jurisdictions nowadays, and in conformity with the TRIPs, in order to be patentable an invention should not only be novel, but it should involve an 'inventive step'. That means intuitively that it should differ from existing techniques in a significant, 'non obvious' way. How and why has such a notion emerged? Although one can find evocation of it in the Venice statute of 1474, it was elaborated progressively by English judges in the sixteenth to eighteenth century and was systematized, in Europe and in the US, in the nineteenth and twentieth centuries. This notion emerged as a response to concerns regarding competition. Requiring simple novelty only means that inventions made as routine business practices, immediate responses to evolution in consumer demand, or 'run of the mills' ideas, could be appropriated by patentees, hence excluding from the market companies which were previously active in the area. The principle was set that no patent should restrain the freedom to operate 'existing trade'. The notion of existing trade includes routine variations of existing products, which a professional of the art would normally find and practice. This intention to protect existing trade is made clear in many judges decisions in Britain. Hence, a concern of economic nature, preserving competition, resulted in technical examination.

Several cases were brought to the Privy Council of Elizabeth I. For instance, in 1571 the company of cutlers attacked a patent that was protecting marginal improvements on knives (haft with bone). 'They used to make knives before, though not with such hafts, and such a light difference or invention should not be cause to restrain them; thereupon he could never have benefit of this patent, although he laboured very greatly herein' (Viner, in Mossoff 2001). In the same line, a 1572 decision states that, 'it is much easier to adde then to invent, and there it was also resolved that no old manufacture in use before can be prohibited' (Coke, in Mossoff 2001). This approach was followed for the next two centuries in England (no patents for mere improvements). Whereas initial practice would allow granting patents to old technology if it had not been worked out in the country for long (hence, no absolute novelty requirement), that changed in the eighteenth century where the courts required absolute novelty. In a decision taken in 1803, Lord Ellenborough mentions that a patent should be granted only,

if the invention be essentially new . . . but if prior to the time of his obtaining a patent, any part of that which is of the substance of the invention has been communicated to the public in the shape of a specification of any other patent, or is a part of the service of the country, so as to be a known thing, in that case he cannot claim the benefit of his patent

(Mossoff 2001)

In France on the other hand, where patents were an instrument of techno-logy policy, aimed at encouraging technological innovation, the focus of exam-ination was technical progress (reducing cost, improving quality of products), not inventive step in its modern sense. However, following the British example, and with the view that the assessment of economic aspects should be left to markets, there was a shift in focus in the 1780s. Certain inventions with eco-nomic value but no significant technical novelty were refused by patent officers such as Berthollet at that time (Hilaire-Perez 2000). With the abolition of prior substantive examination in the 1791 law, even the assessment of technical nov-elty was left to courts.

In addition to regular patents, the French law of 1791 created the 'brevets de perfectionnement', which gave protection over minor improvement of an invention and were used in fact to expropriate the initial inventor by adding only minor features. As such patents were used for merely business stealing strategies, they were restricted in the 1844 law (in which no such patent could be taken less than two years after the initial patent).

In the US, the statute of 1790 mentioned the criterion of 'sufficient importance'. The notion of 'non-obviousness' (roughly equivalent to the European inventive step) emerged progressively in the nineteenth century, following the Supreme Court's decisions, the first of which defined (patentable) invention as, 'something more than the work of a skilled mechanic' (*Hotchkiss v. Greenwood* 1850). In 1941 the Supreme Court again prescribed the 'flash of genius' as the standard of patentability (the *Cuno Engineering v. ADC* case), a notion altogether extremely demanding and vague. The Patent Act passed by Congress in 1952 aimed at superseding this notion that many, notably in the pharmaceutical industry, saw as too stringent (Kingston 2001), replacing it by the weaker requirement that the subject matter should not be, 'obvious to a person having ordinary skills in the art to which it pertains'. This approach was confirmed by decisions of the Supreme Court in the 1960s. Decisions taken by the Court of Appeal of the Federal Circuit in the 1990s have progressively weakened the requirement for non-obviousness, which is now very close in practice to, 'qualified novelty': essentially, an invention which is not explicitly pointed out in the prior art should be considered as non-obvious.

Various approaches were taken in Europe in the twentieth century. UK courts held in 1929 that 'a scintilla of invention' was needed to support the subject matter of a claim (*Parkes v. Cocker*). The German law evolved the notion of technical advance, or 'erfindungshöche'. The Dutch patent office continued to implement a very high standard for invention, resulting in a 10 percent grant rate. When the EPO was formed one of the first steps was to set up a common, European approach regarding inventive step. After much elaboration and discussion, the result was the 'problem and solution approach', which is a method for assessing the inventive step with criteria which are as objective, hence as standardized, as possible. The EPO notion of inventive step is often seen as an average of the UK and Dutch notions, closer to the German practice (Braendly 1973). The inventive step has until now escaped worldwide harmonization. It is part of negotiations of the Substantive Patent Law Treaty conducted at WIPO since 2004. The difficulty is notably that, beyond a formal definition, the notion of inventive step is to a large extent a question of practice, involving the actual way examiners are trained and work: hence its harmonization across separated organizations (even within a single organization) is not straightforward.

DISCLOSURE

Disclosure is now presented in certain patent doctrine as being the counterpart society requests from the patentee in exchange for the exclusive right of exploitation. It is therefore a key component of the patent system. It was not always the case. Disclosure was either absent, or narrowly interpreted, in the early days of the patent system. It was only in the twentieth century that it became central in Europe. In England, until the mid-eighteenth century, no full disclosure was required, even to the authority in charge of examination (just claims were to be submitted, no specification) (Mossoff 2001). A specification section became progressively required after the courts' decision in 1734. Lord Justice Mansfield wrote in 1778, 'you must specify upon record your invention in such a way as shall teach an artist, when your term is out, to make it—and to make it as well as you by your directions: for then at the end of the term, the public have the benefit of it' (Mossoff 2001). This becomes clearer in the Watt case, in which Justice Buller declared, 'the specification is the price which the patentee is to pay for the monopoly'. In England, before 1852, patent specifications were open to the public on payment of a fee, but they were not published or indexed. At the same time (1857) the Patents Museum was opened in London, where patented inventions would be exhibited.

In France no obligation of disclosure to the public was implemented before the end of the eighteenth century. Some patents were published, most were not. A description of the invention was deposited at a notary's office and would be shown only in case of alleged infringement. In fact, even experts in charge of examining applications were often not informed of the actual working of the invention, the inventor would just show them that the invention worked and was more efficient than the competition. However, certain public officials were aware of the usefulness of disclosing new techniques for easing further inventions: an exhibition room and technical library was set up in 1783 by Vaucansson at the Hotel de Mortagne in Paris (Hilaire-Perez 2000). The disclosure requirement is closely linked to the refocusing of examination on technical aspects: in order to assess the technical novelty of an invention, ones needs to know of its technical contents, and to integrate it in some data base of prior art that will be used for examining subsequent applications. Disclosure was made compulsory in the law of 1791, but no effective means were set up for implementing it. The title and abstract of granted patents were published after 1844 in annual catalogues. Inventions seen as strategic for the country would escape this obligation, and it seems that the notion of 'strategic' was quite broadly interpreted. Until 1902, only the manuscript was open at the office of filing, with the visitor having to state his motives and foreigners having to be accompanied by a French attorney, and no extract of the manuscript could be copied before the patent was expired (Galvez-Béhar 2004).

The German law of 1877 prescribed the publication of claims and specification prior to grant. In the US, diffusion of patented knowledge was a concern from the beginning the patent system. Lists of patents were published after 1805, patent description were made available in regional offices of the USPTO in the course of the nineteenth century. The US model has influenced the disclosure policy of patent offices for all the twentieth century. The publication of patent applications and grants is currently a central mission assigned to patent offices, which invest significant resources for handling it by using the latest techniques (now the internet is the major vector of patent information).

RESTRICTIONS TO PATENT RIGHTS

There have been, throughout history, various types of restrictions put on the use of patents, with the view to make them more useful (or less detrimental) to society. The first one is the working requirement, a clause that specifies that the patent holder should implement the protected invention within a certain delay; otherwise the patent could be revoked. Such requirements were put in most patent laws since the beginning, but they have been progressively weakened under international pressure, until the TRIPs of 1994 which allows them only under very narrow conditions.

In England, the Statute of Monopolies prescribed that the patent holder had an obligation to put into practice the patented invention within one year. There is a case in France of a patent granted in 1735 and revoked in 1736 as the holder had not set up a business (Hilaire-Perez 2002). Patents grants in the Ancien Régime frequently included specific obligations regarding exploitation (e.g., the number of factories that the holder should set up. The French law of 1844 gives the following justification, 'It would be injurious to society at large, to allow any one individual to cramp the efforts and attempts of more industrious inventors by obtaining a patent upon which he did not intend to work'. The French law of 1791 included revocation of the patent in case the patentee imported the patented product: this clause was abandoned after the Paris 1884 convention, which ruled it out.

In the German law of 1877, a patent could be revoked after the first three years if it was not effectively put in use, if the owner refused to grant licences or if the invention was primarily exploited outside Germany. In the US (1836 law), foreigners had to exploit their patented invention within 18 months. In Japan (1888), there was a local working requirement of three years.

Working requirements were allowed by the Paris convention, but progressively restricted later: their duration was limited to three years in 1911, and only if the patentee could not justify why the patent was idle. In the TRIPs of 1994, 'local working requirements' are banned—meaning that the working requirement should be worldwide: a patented invention implemented in any one country could not be revoked on this ground only in any other country signatory of the TRIPs.

Compulsory licensing is another possible restriction on the rights of patent holders, as it prescribes them to make their invention available for others to use it, although with fair compensation. In Germany in 1877, compulsory licensing was a preferred solution to revoking patents in case of not fulfilling the working requirement. It was introduced in British patent law in 1883, strengthened in 1919 as 'licences of rights'. Until 1977, these licences of rights enabled British manufacturers to compel foreign patentees to permit the use of pharmaceutical and food products.

2.5. **Major Lessons from History**

Looking at the history of the patent system is enlightening regarding its current status, its rationale, and the underlying forces currently at work. The genesis of major aspects of the system is also revealing of their purpose, sometimes forgotten, buried under immediate concerns.

History reveals the social status—the role and modus operandi—of the patent system: a legal device aimed at encouraging innovation and the diffusion of technology through an economic mechanism. Let's spell out that definition. A legal device means that patents are not a discretionary instrument; they entitle their holder with certain rights as well as obligations, and finally all inventors are potential applicants. Encouraging innovation and diffusion is at the heart of the design of the patent system (the inventive step, the disclosure requirement, etc.) through an economic mechanism, as opposed to an administrative mechanism which would allocate some amount of resources to the inventor: a patent gives its holder the right to market exclusivity, but customers keep a strong influence, and competition still prevails. Only by convincing customers and outperforming competitors can the patent holder realize the potential gains attached to his or her title. If the invention is not of sufficient quality, or if it is too expensive, or if competitors offer even better inventions, customers will not accept to pay a premium price: that option constrains the strategy of the patent holder. Patents are a policy instrument which can be effective only if the government keeps out of the process sufficiently, leaving technical and marketing choices to businesses. It was not always the case. At least in the early days, the temptation was big for the state to interfere more with the inventors, to use the patent system in a more discretionary way. The patent system came to birth with competitive markets and it made progress along with other market institutions. The mercantilist state used market incentives for serving its own interest, increasing the technology pool available to the rulers in competition with other states. Logically, as being the last resort user, state authorities would then decide which techniques were eligible to patents or not according to the contribution of the technology to its power. Progressively, as markets and society became more powerful in front of the king, the rule of law superseded discretionary management of the system. Patents became governed by law and enforced by courts, starting in Britain in the eighteenth century. It also means that, at the very heart of the patent system lays the need to respect competition, to limit market power. This is directly reflected in certain features like the novelty requirement, the limited duration of patents or strict limitation to compulsory licensing.

Within this permanent framework, which defines the identity of the patent system, there is a clear historical trend of structuration and geographical extension of the patent system. By structuration we mean that the law has become broader (covering more configurations with more detail) and that bodies in charge of implementing the law have become stronger: patent offices, specialized courts. In parallel, a 'patent community', a set of patent managers in companies, attorneys, judges, examiners, and other officials emerged. Exceptions to patent law (e.g., drugs) have been reduced, the duration of patents has increased, and limits to the rights of patent holders (compulsory licensing) have been restrained. Geographical extension is also clear, as the number of countries endowed with patent law and patent institutions have permanently increased over the past two centuries. Geographical extension and structuration interact through the process of upward harmonization which has been at work since the late nineteenth century. Simple quantitative analysis confirms this trend: among 271 legal measures on patents taken in 60 countries between 1850 and 2000, two-thirds strengthened patent rights (increased duration, expanded subject matter, cancellation of restrictive clauses such as compulsory licenses, etc.) (Lerner 2002). It is clear also that this upward trend has not been smooth. It has gone through phases of acceleration (the fifteenth century in Italy, the late eighteenth, late nineteenth, and late twentieth centuries) and subsequent slowdown or even some reversal of the process.

What are the forces at work behind this trend and its variations in pace, what forces shape the dynamics of the patent system?

- *Technical change.* All historical phases of waves of technological innovation have experienced strengthening of patent regimes in the concerned regions: the Italian Renaissance, the Industrial Revolution in Britain, the Second Industrial Revolution in Western Europe and the US, and the Information Revolution at the end of the twentieth century. There is a clear correlation between reinforcement of patent systems and acceleration of technical change. Without presuming of whether patents are effective in encouraging inventions, there is certainly causality also the other way round. More inventions, more inventors, more value generated through inventions mean that more value can be extracted from patent protection, giving then incentives to law makers to strengthen patent law. In all the periods mentioned above, policy makers made innovation a clear goal, and patents one means for achieving that goal. More innovation means that the amounts at stake with patents is higher, making it worthwhile for patent holders to claim stronger protection and making it reasonable for society to accept it. Conversely, in phases when economic growth is mainly based on other factors such as the diffusion of existing technology or the accumulation of physical capital (e.g., in the post World War II period) the case for patent protection is weaker and it has few advocates (not many inventors). In certain cases, the strengthening of patent regimes resulted in abuses of the system once the wave of technological innovation started to slowdown— in the form for instance of extortion of revenue by holders of dubious patents, through forced licensing, fraud on financial markets, etc. The patent system had become then too strong in view of the underlying technological opportunities, and it would result in rent seeking strategies. Society would then react by making it more difficult to obtain a patent and monitoring more closely the use of patents (applying more strictly competition law). That happened notably in the mid-nineteenth century after the First industrial revolution (the 'patent controversy').

• *The broader institutional context*, notably importance and degree of development of markets. As an element of the market machinery, the patent system becomes more important when the market is more important. When markets are relatively weak and governments have a leading role (through publicly owned companies, regulation, etc.) technology policies will be implemented first through direct administrative instruments, such as grants, subsidies, or public research. That was the case under the mercantilist state, but also in the post World War II period. It was a time when little patents would be taken. That mechanism works also across countries: countries where the market forces have traditionally been secondary to the state had a relatively weak patent system, as in the case of France. At the end of the twentieth century, as deregulation of markets, privatization of public utilities, and reductions in government R&D expenditure enhanced the role of markets vs. government in various matters including technology, patents regained importance as instruments of technology policy. That was especially clear in the US after 1990, when defence related expenditure on R&D dropped and businesses had to target civilian, competitive markets, they would sharply increase their patenting activity.

• *Internationalization of economic activities*. The reinforcement of patent law in follower countries has always been conducted through pressure from lead countries. That was true in the late nineteenth century (Paris convention) as in the late twentieth century (TRIPs): this pressure was stronger as the lead country had more at stake, meaning more innovation and more of its economic interest related to foreign countries. The latter clearly reflects the degree of internationalization of economic activity: importance of foreign markets and of the circulation of technological information across borders.

- *European construction.* The EPO resulted from the will of European governments to integrate more closely their countries. Current centripetal forces at EPO essentially echo what is going on in the broader European institutions, notably in the EU. Increasingly countries keep to their own immediate interest, sacrificing the broader interest of the group, and thus their own longer term interest.

2.6. **Summary**

- The first patent statute was issued in Venice in 1474. Patents diffused over Europe after the Renaissance. They were initially conceived as 'royal privileges', managed under the discretion of the ruler who would grant such monopolies to inventions or inventors serving his interests or those of the State. A patented invention had to be locally new and effectively put in use.

- Progressively, notably in Britain, the notion emerged that patents should be governed by law and courts. This evolution coincided with the progress of other market institutions and of the modern state. Preventing patent owners from hampering regular market competition was a major concern of the courts.

- The nineteenth century experienced the birth of national patent offices, examining patents prior to grant, and implementing well specified criteria. The twentieth century saw the diffusion of that model worldwide and a trend towards international, upward harmonization of patent regimes, pushed by advanced economies, notably the US, through international treaties, culminating with the TRIPs (Trade Related Intellectual Property Rights) in 1994, part of the WTO founding treaty.

- The European Patent Office (EPO) started its activity in 1978. It has brought a common examination procedure and common standards for patenting all over Europe.

- The central characteristics of modern patent systems include: the subject matter (technical inventions in Europe); priority rule (first to file); inventive step; a statutory duration of 20 years after filing; prior examination of applications before grant; inventive step as minimum degree of novelty; disclosure of the protected invention; and the various but narrow restrictions on patent rights such as compulsory licensing or working requirements.

3 Patents as an Incentive to Innovate

Dominique Guellec

The present chapter focuses on the economic justification and impact of patent systems. It addresses the following questions. What is the justification for society to have patents in the first place? What role do patents play in society? How is the patent system articulated with other policy tools fulfilling similar or complementary roles? What is the economic impact of patents, on patent holders, on third parties, on the economy at large?

In the first section we present the two basic justifications of patent systems—the natural rights and utilitarian theories—the latter being the core of the economic approach. Section 3.2 looks at patents as a policy tool aiming at stimulating innovations. It compares patents with other policy tools having a similar purpose, public research and the public funding of R&D. The third section addresses the impact of patents on innovation. Section 3.4 investigates the broader effects and side effects of patents on the economy.

3.1. The Rationale for Patents

MORAL JUSTIFICATION (OR REJECTION) OF IP

The notion that inventions could be appropriated by man was excluded both by the Romans and by the Catholic Church, on the ground that nature is the creation of God(s). Inventions are just discoveries by man of existing natural properties, created by God(s). Giving ownership to inventors would mean depriving God(s) from their own natural right, which is not acceptable. The natural rights approach to patents takes the opposite position. It is rooted into John Locke's work of the seventeenth century and was pursued more recently by Nozick (1974). It essentially states that an inventor, like any other worker, is entitled 'naturally' to own the result of his or her work. More precisely: a person who labours upon resources that are either unowned or 'held in common' has a natural property right to the fruits of his or her labour. Locke states in his Second treatise of Civil Government (1690):

Though the Earth, and all inferior Creatures be comon to all Men, yet every Man has a Property in his own Person. This no Body has any Right to but himself. The Labour of his Body, and the Work of his Hands, we may say, are properly his. Whatsoever then he removes out of the State that Nature hath provided, and left it in, he hath mixed his Labour with, and joyned to it something that is his own, and thereby makes it his Property. It being by him removed from the common state Nature placed it in, it hath by this labour something annexed to it, that excludes the common right of other Men. For this Labour being the unquestionable Property of the Labourer, no Man but he can have a right to what that is once joined to, at least where there is enough, and as good left in common for others.

That view, endorsed by some philosophers of the Enlightenment inspired the first patent law in France in 1791, which states that, 'Every discovery or invention, in every type of industry, is the property of its creator; the law therefore guarantees him its full and entire enjoyment'. The French law of 1791 establishes that the same principles regarding property or land should apply to ideas. That was used as an argument in favour of a registration system for patents (without examination): if it is a natural right, then government should not interfere with it, it should simply ensure its enforcement. The natural right approach puts patents in the realm of law only, outside the scope of policy. It is then consistent with an exclusive monitoring of the system by courts, based on legal principles and not on policy objectives. This approach also sees patents as a reward, fairly deserved by the inventor, rather than an incentive serving society's interest.

This argument was put forward by Locke regarding land and agriculture: it would notably apply to new British settlements in America. When applied to intangible assets it is not straightforward, however. In fact, it requires refinement rooted into deep and strong statements regarding the nature of inventive activities. Hence a proviso that Locke himself mentioned regarding the validity of his vision: that 'there is enough and as good left in common from others'. One problem with the natural rights approach applied to inventions is that one person's right is another person's exclusion while it is in general unclear to what extent one invention can be attributed to one inventor. 'Only God creates from scratch' as a US Judge famously said. In the late nineteenth century Joseph Lewis Ricardo, founder of the Electric Telegraph Co, argued that since, 'nearly all useful inventions depend less on any individual than on the progress of society', there is no need for it to, 'reward him who might be lucky enough to be the first on the thing [invention] required'. The argument here is that the world of inventions belongs to society, with the same effect as when it belonged to God(s)—it should not be appropriated by any individual human being.

Everyone's invention is based on accumulated knowledge, the sum of past inventions made by others. This is the 'Commons' referred to by Locke in that context. Granting anyone control over the latest invention endows him or her with a de facto control over previous inventions that the new one is based on and displaces from use, as it is presumably superior. This is a backward looking counter argument. There is also a forward looking equivalent: by granting a right on a current invention, one deprives possible future inventors of that particular right. If a given person had not made the invention now, another person would have probably done it later, and this person would also be entitled legitimately, in that context, to claim property over the invention.

It could also be argued that inventions come by chance. A famous decision by US Judge Arnold, in 1941, reads,

Each man is given a section of the hay to search. The man who finds the needle shows no more 'genius' or no more ability than the others who are searching different portions of the haystack. The 'inventor' is paid only a salary, he gets no royalties, he has no property rights in the improvements which he helps to create. To give patents for such routine experimentation on a vast scale is to use the patent law to reward capital investment, and create monopolies for corporate organizers instead of men of inventive genius.

(Schmookler 1966: 31)

Box 3.1 on '*Diderot v. Condorcet*' further illustrates the opposite views on the intellectual property rights.

The argument that man is naturally entitled to own ideas can simply be opposed the opposite statement: ideas are naturally free of any ownership. As Jefferson put it in 1813 (in his letter to Isaac McPherson):

That ideas should freely spread from one to another over the globe, for the moral and mutual instruction of man, and improvement of his condition, seems to have been peculiarly and benevolently designed by nature, when she made them, like fire, expansible over all space, without lessening their density in any point, and like the air in which we breathe, move, and have our physical being, incapable of confinement or exclusive appropriation. Inventions then cannot, in nature, be a subject of property. Society may give an exclusive right to the profits arising from them, as an encouragement to men to pursue ideas which may produce utility, but this may or may not be done, according to the will and convenience of the society, without claim or complaint from anybody . . .

The natural rights approach justifies the differentiated treatment by patent law of inventions and discoveries. An invention is a creation by man, whereas a discovery pre-existed to its finding. The first notion gives a more important role to the inventor, it ties the finding to the finder, the finding is as unique as the finder is. America would not be different if its discoverer (for Europeans) had not been Cristobal Colon but some other explorer. Conversely, electrical appliances could have followed a slightly different path without Thomas Edison. Extreme cases are to be found in arts, where, say, *A la recherche du temps perdu* would not have been written if not by Marcel Proust. Taking a natural rights approach would lead to give a higher reward to Proust as to Edison, to Edison as to Colon, on the basis of 'merit'.

THE UTILITARIAN APPROACH

The utilitarian approach claims that social institutions should be designed so as to maximize social welfare. The core of the utilitarian argument for patents is that free competition will generate an under-optimal rate of inventions, due to the 'public good' characteristic of knowledge. Hence it is the interest of society to supplement free competition with special institutions in that field, patents being one of them. The utilitarian approach views patents as incentives for further innovation, not as rewards for past innovation: the two notions are very similar in cases where the use of patents would be recommended by the utilitarian approach, but this recommendation is not made in certain cases as there is no need for further incentives than the ones provided by markets or governments. The utilitarian approach views patents as a policy instrument, tied to certain aims and circumstances. The questions this approach leads to are then: in which circumstances should this instrument be used? How to design this instrument so that it fulfils its mission in a satisfactory way?

The starting point for the utilitarian approach is similar to Jefferson's argument quoted above: a given piece of knowledge can be used at the same time by different persons in different places, and does not disappear from its use. That differentiates knowledge from tangible goods. An apple can be eaten by one person only at a time, and disappears once it has been eaten. That peculiar property of knowledge is named 'non-rivalry'. It makes knowledge a 'public good'. There is no congestion in the use of knowledge goods, which makes them even more public than other public goods such as roads. That property has consequences on the economy of knowledge.

First, the marginal cost of using a piece of knowledge is zero. Hence the cost of invention is a sunk cost, to be incurred once before the production of the invented good, which has variable costs, starts.

Second, re-inventing an existing piece of knowledge is a waste of social resources. That occurs for instance when a company duplicates a product first invented by a competitor. Once an invention is made, it is beneficial to society that it is made available for free to all potential users: As the cost of the invention has been incurred already and its use generates no further cost to the inventor or to other users (the opportunity cost is zero), unlimited and free access is socially preferable.

Third, an existing piece of knowledge can be beneficial to others than its inventor, without them needing to incur again the cost of invention and without depriving the inventor of the use of his invention. This is a 'positive spillover'. It implies that the social return on inventions is usually higher than the private return, as only part of the return accrues to the inventor.

Fourth, as the private return is lower than the social return, certain inventions whose social return would justify the expenditure needed to obtain them will not be made due to insufficient private return. The owner appropriates only part of the benefits generated by his invention, and that part can be too small to justify the investment. Hence the competitive market mechanisms might not generate as many inventions as society would be willing to have.

Fifth, a competitive market could make things even worse, as an inventor must charge a price that will allow him to recoup his fixed cost while his competitors/imitators can charge just their marginal cost, hence driving the inventor out of business. Anticipating that situation, companies will not invest in research in the first place.

In fact, the market mechanism would lead to under-investment in research in general, but also possibly to excessive investment in particular areas: in the absence of legal protection companies would keep their inventions secret, making it necessary for others to duplicate their research projects—which is a waste of resources.

These various economic attributes of knowledge call for intervention by government, while putting conditions on how such an intervention should be designed. The problem for society is to find mechanisms which allow the channelling of resources to knowledge producing activities.

One solution for government is to sponsor inventors or inventions, and to make inventions free to all users. Inventions can be put into the public domain so that any company could use them for free and, under pressure from competitors, could not charge customers for the research cost. Public sponsorship is the traditional response of economics for the provision of public goods (Samuelson 1954).

The symmetric solution is to 'privatize' knowledge, to make it an 'excludable good', in other words, to erect a mechanism that excludes others than the inventor from using the invention. This is intellectual property rights (IPR). The holder of IPR can then either keep exclusivity of the invention or give others access to it under conditions that make it economically rewarding to them. In any case, this exclusive right can translate into an extra-reward, beyond the normal competitive profit, that could allow recouping the cost of research and compensating for the risk.

By granting exclusive rights, society is also generating costs, as these rights hamper the access to existing inventions, hence reducing positive knowledge spillovers. The utilitarian view of patents has at its core a trade-off between benefits (incentive to invent) and costs (reduced diffusion). The balance between benefits and costs could varies upon different circumstances, which must be assessed in each case.

A variation of the utilitarian argument is often put forward, which has become standard legal view of patents nowadays in Europe: it states that patents are a contract between the inventor and society, by which society grants transitory monopoly to the inventor in exchange for disclosure. Patent is here a response to secrecy, not to under-investment. This theory emerged progressively from British judges in the eighteenth century. It reflects closely one of the requirements enforced by patent offices, namely that the invention should be disclosed in the patent document. Although it has some truth, this theory is incomplete, it can hardly give account of the current state of patent systems. First, it takes the invention as given: it is true that once the invention is made, one major objective for society is to have it disclosed. But that comes after the invention is made, and having the invention made in the first place is the primary objective for society. For that purpose, patents operate as a very special type of 'contract', or a promise: society commits, ex-ante, vis-à-vis inventors, to grant them some exclusive right if they come with an invention. That is the primary goal of the patent system, disclosure comes second only. Second, reducing patents to a contract over disclosure would result in the recommendation to grant patents only to those inventions that could be kept secret by the inventor if they were not patented. Otherwise society has no benefit in the contract. That would exclude for instance most product innovations, which consumers and competitors can immediately grasp when buying a new product. This does not correspond to the reality of patent law, which does not treat differently product and process innovations (one exception could be drugs, for which certain countries used to allow process patents only, not product patents; the rationale being not to favour disclosure, but to favour competition as the product would remain in the public domain). Overall, disclosure is certainly one objective of the patent system, but it comes after the provision of incentives to invent.

ARE PATENTS PROPERTY RIGHTS?

When addressing the utilitarian foundation of IPR, one is led inevitably to compare it with tangible property: to what extent, if any, does the economic theory of tangible property apply to intangible goods? Full application is defended by some (Duffy 2004), on the ground that both address the issue of internalising externalities and promoting investment in a market framework. What is at stake here is essentially the strength to be given to patents: If they are seen as property rights, many bodies of law providing strict defence of property would apply, whereas keeping them out of that domain gives more flexibility.

The economic theory of property rights has been developed by Demsetz (1968), and it could be traced back to David Hume in the eighteenth century, who, in contrast to Locke, built a utilitarian theory of property rights. It grounds property rights on the fact that they allow internalisation of externalities, hence promoting social welfare. Assets which are not subject to private property are subject to over-exploitation—it is the 'tragedy of the commons'. This is exemplified by fishing pools: each individual fisher has interest in capturing as many fishes as he or she can, although the aggregate result of individual actions is the depletion of the pool, which inflicts harm to all the community. This is individually rationale behaviour, as each fishers knows, if they restrain themselves, it would be in the interest of each other fisher to capture what they have left instead of leaving it for the repletion of the pool. Each individual's actions inflict a cost to the community, a negative externality, which creates a 'social bad'. Property rights provide a solution to that problem. If the pool were to have a single owner, he or she would have interest in maintaining it for future use as he or she will get the reward accruing from that use. To what extent does this argument, which was put forward for tangible assets, also work for intangible ones? Marc Lemley (2004) argues that it does not hold, as Demsetz's theory is about internalizing *negative* externalities (which should be suppressed), while knowledge is associated with *positive* externalities (which should be encouraged). Therefore, internalization should be partial only for intangibles and rights of control, weaker than in the case of tangible assets, should be allocated accordingly. Duffy (2004) rejects that view, arguing that what is a positive externality for some (the users of knowledge) is at the same time a negative externality for others (producers of knowledge). There is no reason to treat intangibles in a different way as tangibles, and the theory of property rights applies as well.

The latter argument is largely wrong from the point of view of economics as it ignores the public good property of knowledge. A positive externality means that, after a transaction (e.g., the sale of a new good) has occurred and was beneficial to all parties involved, further value is created which will accrue either to one of the participants or to third parties. A positive externality means net creation of value, while a negative externality means net destruction: the difference between the two is not a matter of distribution of value, as argued by Duffy, but of creation vs. destruction. Rights of control should be designed so as to facilitate beneficial transactions, and society has interest in the distribution of value between the parties only to the extent that it will affect the possibility that the transaction occurs or not, or the total value created through that transaction. If a transaction would take place anyway, in other words, if the invention would be done and diffused without patents, all what a patent could do is to reduce the social benefit (the externality), something that society has no interest to favour. Any intervention in that case could be justified only on distributive grounds— society wants to increase the share of revenue accruing to certain parties at the expense of others— not on efficiency grounds.

However the economics of property right is still relevant for intangibles, although in a dynamic way. Let us go back to the traditional argument raised by Plant (1934). Plant stated that patents could not be justified as property rights, as the latter aim at reducing shortages (e.g., of fish in the example above), while patents instead create shortage, by reducing the access of society to existing technology. 'Property rights in patents and copyrights make possible the creation of a scarcity of the products appropriated which could not otherwise be maintained . . . The beneficiary is made the owner of the entire supply of the product for which there may be no easily obtainable substitute'. This is a static view, in the sense that it applies ex-post only, to existing inventions. When taking a dynamic view, the question becomes: how to encourage new inventions to occur, now and in the future? In that dynamic setting the role of patents is to reduce a shortage: but the shortage of new technology, not the access to existing technology. That is the specificity of IPR as opposed to property rights on tangible assets: whereas the latter serve the purpose of managing current scarcity of resources, the former aim at reducing (future) scarcity by inducing more investment. Partial and temporary scarcity ex post is a compromise which aims at solving absolute scarcity ex-ante.

Finally, the economic view of property rights provides further insights into the rationale for patents. The theory developed by Hart and Moore (1990) emphasizes the role of property rights as incentives. The stand point is incomplete contracts. Contracts can seldom be complete, because of the limitations of individuals' rationality, availability of information etc. That is damaging because of possible opportunistic behaviours by some parties. All the contingencies cannot be anticipated and set into the contract so that there is ample space for ex post disagreement. In this theory, the owners of an asset have the residual rights of control on that asset: they decide what to do with the asset once certain obligations contracted with third parties (e.g., banks and workers) have been met. After the contractual rights of other parties have been exhausted at predefined conditions (e.g., interest rate for loans and wages for workers), the owners of the asset keep what revenue is left. The owners are the residual claimants on the income generated by the asset. Property rights provide high-powered incentives to invest and produce value, as they tie together residual rights of control and residual revenue. The decision maker bears the full financial impact of his or her own choices and is expected to act accordingly. Hence a patent is expected not only to increase the share of income accruing to the inventor, but also to induce him or her to use the invention so as to maximize the value it generates. This approach gives account of the broader use of patents that has been developing over the recent past. Patents are not only a means for their holder to extract more value, but they have become a way to access others' technology (licensing) or to raise capital (signalling). In accordance with that view, patents give a stronger incentive to their holder to commercialize the invention, as the revenue increases in the quantity sold (and the price charged) of the commercialized product embodying the invention. It is on the basis of that principle that government have encouraged since the 1980s universities to patent their inventions and to grant licences to businesses which then commercialize the resulting product, with a presumably stronger motivation than if they had no guarantee of exclusivity. The right of control gives to its holders the possibility and the incentive to conduct various types of actions that will generate value for themselves and, in many cases, for society.

At times of expanding and strengthening patents, it seems that the '*natural rights*' argument and its offspring tend to have more favour in the courts. Lemley (2004) shows that the notion of property right has been increasingly invoked by the courts in the US over the past two decades. The reason is that the natural right argument just refers to a principle, with no need of further substantiation, whereas the utilitarian argument calls for empirical proofs—it supports patents only in certain circumstances, when they increase social welfare, and it must be checked in each case that the proper conditions are met. The natural right is therefore an easier argument to manipulate. The natural right approach also keeps patents in the realm of rights and courts, out of the realm of policy and considerations that are not purely legal. At the opposite, the economics of patents can be considered as an elaboration of the utilitarian theory of patents, taking into account, testing, and drawing implications from further assumptions regarding the behaviour of inventors and customers, the characteristics of inventions, the informational framework, the market structure and the objectives of government.

Box 3.1. *Diderot v. Condorcet* (based on Hilaire-Perez 2002)

The controversy between two French philosophers of the Enlightenment of the eighteenth century, Diderot and Condorcet, illustrates the opposition of the natural rights and the utilitarian visions. Diderot was a strong supporter of 'droits d'auteur' (copyright) as a natural right for writers and artists, as they are real creators. On the other hand, he viewed new inventions as resulting from the combination of existing inventions, consisting in just establishing relationships between known facts or natural laws. Combination is not creation: it is a matter of method and chance, not of genius, contrary to the arts. Any person having ordinary skills in the art and devoting enough time and resources to the matter could come with a combination having the desired properties. In addition, invention is a collective process—many researchers are searching the same area at the same time, and the first to find is the luckiest of all, not necessarily the best. Hence no *personal* merit such as a patent should go to inventors. Not only patents are not morally grounded, but they have a detrimental effect on invention. Patents erect obstacles in the process, as combinations can be made only if the knowledge can be easily accessed and used. Of course, if they cannot protect their inventions by patents, inventors will keep them secret whenever possible: in order to prevent that, invention activities should be sponsored by government, which would ensure full publicity of inventions. Condorcet, himself a scientist and mathematician, had an opposite view to Diderot. He favoured patent protection for technological inventions, in order to increase the return to the inventor by mitigating imitation, while he was sceptical with copyright for artistic and intellectual creations. Artists and intellectuals have an inner motivation which does not need to be supplemented by financial incentives, and such exclusive rights put obstacles to free, democratic discussion.

3.2. **Patents as Policy Tool**

Encouraging innovation is a goal that all governments have on their agenda. Markets do not generate a socially efficient level of innovation on their own, as knowledge has many features of a public good, and as investment in innovation is in general more risky than other types of investment, deterring capital from entering that area. A complementary goal to fostering innovation is encouraging the diffusion and use of new technology, as it is the only way to enhance durable productivity growth and to generate further knowledge. Governments have a range of instruments for encouraging innovation and diffusion of technology, including patents. What are these instruments? How do patents compare to them? How do patents fit with them?

TECHNOLOGY POLICY

Major policy instruments for encouraging invention fall into three broad categories. The first category is the public research system: it is made of universities and of public laboratories, controlled by the government. The second instrument is public funding of business performed research: subsidies (targeted or not targeted), prizes, soft loans, or tax incentives. The third is intellectual property policies in general and more particularly patents. The three instruments have very different properties from an economic point of view, regarding how they allocate funding to inventive activities, the way technical and economic decisions are taken, and the required informational structure for them to work. Each instrument has its advantages and drawbacks, which may feature more or less prominently depending on the context and the precise objectives of government.

Table 3.1. R&D expenditure of OECD countries (2003)

	Europe	US	Japan	OECD
GERD: Global R&D expenditure (billion USD)	211.3	292.4	112.7	686.6
GERD as % of GDP	1.8	2.7	3.2	2.3
Business performed R&D (BERD) as % of GERD	63.3	69.8	75.0	67.7
R&D funded by government as % of GERD	35.5	30.8	17.7	30.4
BERD funded by government as % of BERD	8.2	10.2	0.8	7.4

The public research system is mainly funded by government, from tax revenue. Ideally, its research areas are fixed by government, on the basis of public needs. Its research covers three broad areas: fundamental knowledge (which has no direct economic use); technology fulfilling collective needs of citizens (defence, space, health); and generic, industrial technology that government is considered better equipped than business to work on (e.g., due to high cross-fertilization with other areas covered by public research). The funding is not conditioned in general by the outcome (success or failure) of the research. However, with budgets being re-allocated on a regular basis, poorly performing teams would suffer in the next budget rounds.

There are two main types of funding mechanisms of public research institutions: grants allocated on a competitive basis to particular projects, following a call for tender by governmental agencies (e.g., the Research Framework Programs sponsored by the European Commission), generally selected by a panel of experts, then there are public laboratories, like the Commissariat à l'Energie Atomique (CEA) in France, which rely mainly on lump sum funding—research teams receive basic funding and their research agenda. In both cases, the funding is supposed to cover the cost of doing the research, not to reflect directly the value of the outcome. This is fair as the government is both funder and user of the research, hence all profit (difference between cost and value) would be simply transfers within the government budget. Research can be controlled by a hierarchical mechanism (orders cascading from the government to individual teams), as it is generally the case in applied research responding to government needs, or it can be largely left to the initiative of individual teams (more the case for fundamental research performed in academic institutions).

The second type of support is the public funding of research performed by enterprises. It involves a transfer of funds to companies, as a counterpart for some research being done. The major mechanisms used by government to transfer funds to private parties are as follows:

- *Public procurement* involves the government purchasing research from a private party; the payment has a transfer of property rights as a counterpart, which means that the results of the research, its intellectual property, belong to the government. A call for tender is often requested, in order to ensure a competitive allocation.

- *Research subsidies* imply the government sponsoring research projects performed by private parties for their own use (no transfer of property) although in general some of the results should also be made available to the public. Such subsidies are targeted, they generally respond to a particular objective (e.g., enhancement of car safety or reduction in the pollution of water); they can be allocated on a competitive basis (through a call for tender, which is increasingly the case) or by simple administrative decision (often going to a consortium of firms so as to avoid criticisms of biased allocation profiles and potential distortion of competition). Research subsidies are limited to a certain share of the cost of particular projects by the WTO, on fair trade grounds.

- *Prizes* are government controlled competitions regarding well defined innovative projects. The prize goes to the first to submit the requested invention, not a research project. This method was used quite often in the eighteenth century, notably in France. It has been revived recently in several instances in the US (e.g., the US$10 million prize offered by the NASA to inventors of robot vehicles succeeding certain tests).

- *Soft loans* can be characterized by a reduced interest rate (thanks to a government subsidy), or a guarantee of reimbursement by the government, or a clause of reimbursement only in case of success. Such loans have been used for instance by the European governments in order to support Airbus (while Boeing benefitted from procurement by the US government). These loans can also be allocated to smaller projects, often conducted by small firms, responding to particular criteria in terms of innovativeness or risk (e.g., loans granted by the ANVAR, a governmental agency in France).

- Finally, public funding can take the form of *tax breaks*. In 2006, 'R&D tax credits' were practised by more than 20 of the 30 OECD member countries, including most EU member states. This policy tool is more recent than the others: it appeared in Canada in the 1960s and only about ten OECD countries used it in 1995. The company benefits from reduced taxation on its profits in proportion of its research expenditure or of the change in its research expenditure over some reference period. The company decides itself on the orientation and level of funding of its projects; they will still be eligible to the tax relief provided that they consist in real research.

These various types of public funding have different economic characteristics. The level of funding is based in most of them on the cost of the research, except for prizes where it is value. In the case of public procurement, the dominant practice is the so-called 'cost plus' method in which government refunds all costs and adds some fair profit margin to the company. Both in the case of cost-based and value-based funding, the issue is the ability of government to identify, or elicit revelation of, the true value or cost of the invention. This informational issue is the major difficulty that affects such policies.

The patent system is the third type of instrument of technology policy. Patents can be characterized in a similar cost/value framework as the two other types of instruments. The exclusive right they give allows the holder to charge customers with a mark up above the marginal cost. Hence the patent system generates a kind of *targeted tax*: It is a tax as it relies on a monopoly granted by government, from which no buyer of the good can escape, and it is targeted as only buyers of the protected good (not all citizens) are charged. That distinguishes patents from other instruments, which are funded through the general tax system, contributed by all citizens. With patents, the funding that accrues to the inventor is closely related to the value of the invention (volume of sales and willingness of customers to pay a higher price, the 'patent premium'), not to the cost of doing the research. The research is generally of the applied, type, as a patent is granted under the condition of industrial applicability of the invention. All economic and technical decisions are taken by the company, not by the government. Related to that aspect is the non-discretionary character of patents. All inventions fulfilling certain criteria written in law are eligible to patents, whereas for most other instruments a case by case decision by administrative authorities is taken—the government can decide to support one particular industry or technology while ignoring others. On the other hand, patents have an exclusionary effect (reduced competition) that other instruments don't have. Although some of these instruments might in certain cases impede the diffusion of technology (e.g., allowing the recipient of funds to keep the resulting knowledge secret), this would be a voluntary and specific move, while the very principle of patents is exclusion. It must be noticed that one financial instrument, the R&D tax credit, also avoids the government to be involved in project selection: all projects fulfilling certain criteria defined ex ante can enjoy the financial support.

While the first two types of instruments (except for procurement) affect the business sector mainly through a reduction of its research cost (through the provision of funds or of knowledge produced by public institutions), the patent system operates through an increase in the value of the research outcome. Hence the patent system intervenes quite downstream in the value chain.

Accordingly, as compared with other instruments of technology policy, patents play the role of encouraging not only research, but also and possibly even more the commercialization of inventions. That aspect is made clear by two features of patent law:

- A patent will generate income to its owner only if a product or process is economically used out of it. So, it is not invention per se which is rewarded, but the potential to disseminate it. The higher the scale of production, the higher the benefit accruing to the patent holder.

- Patent infringement is punished only when identified by the owner itself. It is not, like certain other infractions, monitored by government. Hence, the likelihood of enforcement increases with the resources that the owner, not the government, invests in monitoring. In turn, these resources are likely to be an increasing function of the private value of the patent. If the market is not monitored by the holder, then imitation can flourish and society will benefit from the invention, which is then actually put into use, at a lower cost.

The respective characteristics of the three types of instruments are summarized in Table 3.2.

Table 3.2. Characteristics of various instruments of innovation policy

	Source of funding	Reference for funding	Type of research	Project selection	Control over results
Public research system					
Public labs	General tax	Cost	Basic, generic	Government	Government
Academic research	General tax	Cost	Basic, generic	Government/ universities	Government/ public domain
Public funding of business research					
Procurements	General tax	Cost	Generic, applied	Government	Government
Subsidies	General tax	Cost	Generic, applied	Company	Company
Prizes	General tax	Value	Generic	Goverment	Goverment
Soft loans	General tax	Cost	Applied	Company/ government	Company
R&D tax credits	General tax	Cost	All	Company	Company
Patents	Customers	Value	Applied	Company	Company

WHICH INSTRUMENT SHOULD BE USED IN WHICH CIRCUMSTANCES?

We define the efficiency of an instrument as its ability to generate more (or higher value) innovation at the lowest cost for society. This is a utilitarian approach. Considerations relating to fairness and distributive impact will also be taken into consideration. The relative efficiency of the various instruments depends directly upon the technical and informational context, which might differ across technology fields or across markets. We focus in what follows on the alternative between patents and other instruments.

The first issue is whether the technology renders direct services to customers (including government) or not, in other words, to what extent the research is applied vs. fundamental. Research areas with no predictable market application will barely attract private funding as no profit is to be expected, with or without patents. Therefore public funding is needed. It could also be the case that applications are expected, but in the long-term only, so that patents (with their limited term) are less effective as a patent might have lapsed at the time the market stage is reached. The issue is raised for instance in nanotechnology, for which major applications are foreseen in the long-term only while much research is being done now already. On the other hand, the notion of application is not so clear-cut. Applications can be indirect, with certain inventions with no direct application serving to develop inventions which have a market— for example research tools. In that case, a market can emerge upstream, for the inventions themselves as intermediate inputs, creating a rationale for patents. However in that case the exclusionary effect of patents could be detrimental to basic research, whose progress requires a complete access to all knowledge available, which is more straightforward in an environment of public funding. There is here a series of trade-offs that are addressed in other parts of the book.

The second issue is the existence or not of substitutes to the invention. Certain techniques might not have efficient substitutes, so that the market power granted by patents is extremely strong. Products which are necessary to customers and for which there is no substitute (e.g., certain drugs) are called 'essential facilities'. An economic doctrine has developed for these particular goods, which emphasizes the need for close control by the government, because private (absolute) monopolies would have detrimental effects on society. One occasion when the essential facility doctrine was (notionally) applied is the threat by the US government in October 2001 to issue a compulsory licence on an anti-anthrax drug that was sold by Bayer at a price that the US government deemed too high.

A third issue is exclusion, which relates both to efficiency and to equity. Patents exclude customers which are not ready, or able, to pay the higher price charged on patented goods. One has to see what is the cost to society of such an exclusion and how this cost is shared among customers. In terms of efficiency, one has to distinguish between two categories of excluded customers: those who would be ready to pay more than the marginal cost of production (but less than the price charged by the patent holder), and those who would not be ready to pay that price. Provided that the R&D cost could be shared among customers ready to pay for the higher price (including the patent premium), it is socially inefficient to exclude customers willing to pay more than the marginal cost, as they do not impose further cost on society while increasing their own well-being. For those not ready to pay even for the marginal cost, the issue is essentially about fairness, and it is not different for patented goods as it is for any other type of good (e.g., food).

There is a further question, related to equity: to what extent is it fair to make all citizens pay for inventions that many of them will not use, as it happens with tax funded instruments? The answer to that question depends notably on the type of service that will result from the research. Requesting all citizens to pay for R&D performed, say for improving comfort in luxury cars, is probably not improving equity in society. On the other hand, requesting all citizens to pay for pharmaceutical research seems more equitable, because all citizens are potential users, and illness does not depend directly on an individual choice. Complementary mechanisms to patents are sometimes used for mitigating exclusion. In the case of drugs, in developed countries, patents are not the only factor affecting prices: in fact, prices are negotiated between pharmaceutical companies and governments (as governmental mechanisms like social security reimburse drugs), with the view to limit the exercise of the market power given by patents and reduce the exclusionary effect that higher prices would have or the cost to social security bodies. In addition, much of the upstream research in the field of bio sciences is conducted by publicly funded teams—the largest R&D budget in the world is the NIH, the National Institute for Health, in the US, with a budget of US$28.8 billion in 2005.

The exclusion effect is certainly stronger for patents as for other types of policy instruments, but those instruments are not immune from it, although in an indirect way. The alternative to patents is not 'free research'. It is government funding, meaning that the cost is incurred by tax payers. The opportunity cost of taxes is not zero. It is less visible as it consists in a reduced level of overall consumption, affecting many types of existing goods and nonexistent ones (which did not emerge as total demand was not high enough). Hence, funding by taxes has also some kind of exclusionary effect, but more diffused. When analysed from a general equilibrium point of view, the exclusion effect of tax funded instruments becomes apparent. It is true however that the exclusion generated by patents concentrates on knowledge intensive goods, whose marginal cost of production is relatively lower, hence having a more detrimental effect in terms of social efficiency than tax funded instruments.

The choice of an instrument is also related to the allocation of information, incentives and decision rights among economic agents. It is desirable from the point of view of efficiency that decisions are taken on the basis of the best information available: hence decision makers should be those who have the information. The relevant information here is the cost and value of the research. Generally speaking, one would expect that the less government knows, the more decision making should go to business, provided that the incentive framework is compatible with social welfare. Proper incentives should lead to minimize costs and maximize value. On that basis, one would not expect for instance a prize system to be used when government has no idea of the value of the invention, or a cost-plus funding mechanism to be used if government cannot check the cost of research. As reported in Table 3.2, the control exercised by government is total for public research, partial for public funding of private research, and nil for patents. A more detailed analysis is given in Box 3.4. The major conclusion is that patents are more efficient, from an informational point of view, when the value of the invention is not known by government: in fact, patents could be regarded as prizes whose amount is set by the market, dependant on demand and on marketing efforts by the supplier. Patents are more efficient as well when, in addition to the value, the cost itself is not observed by government.

The different instruments are not necessarily exclusive of each other. Government can use several instruments at the same time in order to address the different difficulties of certain areas or types of research. A prominent

---

BOX 3.2. The relative efficiency of policy instruments in various informational conditions (based on Scotchmer 2001)

Assume an economy with a government representing the interest of society and a large number of separate entities which have the ability of generating inventions, either businesses or public research labs. The socially optimal situation is such that (1) the research is done only if its social value is higher than its cost; and (2) in that case one entity only does the research—the one with the lowest cost. The question is, What policy mechanism will most likely lead to that situation (i.e., ensure that desirable inventions are generated at the lowest cost)? Three cases are distinguished on the basis of the availability of information to government and the research labs.

1. Government knows the value of the invention and the cost of various labs (or at least it knows those for which the value is higher than the cost). As all relevant information is known or observable, the solution is straightforward. The government can allocate the research to the most efficient laboratory and pay just for its cost, then putting the invention in the public domain. So a subsidy is the preferred mechanism.

2. Government knows the value of the invention but not the cost, and individual laboratories know both (each knows its own cost but not others' cost). A subsidy will not work as it needs the cost to be known by the government. A prize equal to the social value of the invention will elicit research from firms whose cost is under that value and have reasonable expectation to win. There are two drawbacks however. First, the prize might be too high, as all of the social value will be appropriated by the winner (nothing left to the rest of society). Second, there might be many competitors, although one (the one with the lowest cost) would suffice, the rest being a waste of resources (Loury 1979).

3. Government knows the cost, or could observe it if needed, but not the value. Firms know the value, but why would they reveal it? Economists have imagined sophisticated mechanisms by which all firms would have interest to reveal the true value, but such mechanisms are not really operational (they generally require extremely sophisticated strategies of players and dense exchanges of information so that each of them integrates all information generated by others).

   In fact, the patent mechanism is probably the best answer. The private value of a patented invention reflects to a large extent its social value. Therefore commercialization of the patented invention will reveal the value, and it acts as an incentive to companies. The same result holds if the government does not know even the cost, in which case there is no alternative mechanism to patents.

---

example is the authorization given to public research entities to file patents. This approach has been pioneered by the Bayh-Dole Act taken in the United States in 1980, now emulated by nearly all other governments in the OECD and beyond. The rationale for doing so is that for certain types of research (1) the knowledge produced is fundamental, so that public labs are better equipped to do the research; while (2) for applications to arise complementary investment is needed that businesses are in a better position to do, but won't do if they don't have an exclusive right. Other examples include the IPR policy implemented as part of the EC Framework Programmes (participants to such programmes benefit from subsidies and are encouraged to file for patents).

Patents are the most market oriented among instruments of innovation policy. The patent system is essentially an attempt by government to create conditions for the market mechanism to work for the production and diffusion of technology. It aims at decentralizing a socially desirable situation (i.e., to leave decision making to individual agents instead of to a central planner) for the production and diffusion of a *public good* subject to *incomplete information.*

As compared with other instruments of technology policy, the patent system aims at offering the advantages of the market mechanism relative to political and administrative processes, in terms of information, incentives, and competition. The signals generated by markets are prices, reflecting the orientation of demand and its response to supply. When markets work well, customers influence patterns of technological change. Incentives are given to businesses to respond to prices by orienting their research efforts accordingly. Companies are also given incentives to maximize the value of their technology hence its use, while limiting its cost. Competition keeps firms under check, so that they make their best to respond to demand and reduce cost. These market outcomes could not be obtained by using alternative policy instruments. It seems paradoxical at first glance to claim that a system of exclusive rights might create competition, although it actually can. The size and sophistication of the patent system, hence its cost, are quite high, due to the complexity of the issues to be solved (e.g., public good, incomplete information). As compared with other instruments of technology policy, however, it is not clear that the patent system is more costly to administer—think of all the panels and administrative employees needed to run public laboratories or competitive allocation systems.

3.3. **An Economic Incentive**

The primary mission assigned to patents by society is to encourage investment in invention (R&D) so as to foster new products and new production processes.

The effectiveness of patents for increasing R&D expenditure at the firm level rests on a series of factors that can be put together conveniently in a model of sequential choice with two stages (Arora et al. 2001; Duguet and Lelarge 2005). At Stage 1 the firm decides whether to invest in R&D or not. At Stage 2, having the invention, the firm decides whether to patent it or not. The model, illustrated in Box 3.6, must be solved in a recursive manner, starting at Stage 2. Faced with the decision to patent or not its invention, the firm compares its expected profit under the two scenarios. Let's call *PN* the expected profit in case no patent is taken and PB the expected profit in case a patent is taken. The net profit in this case is PB-C, where C is the cost of taking a patent. (the cost of drafting the application, administrative and legal fees, and the cost of enforcing the patent). The expected profit for the profit maximizing firm is:

$$\Pi = Max\ (PB-C, PN).$$

Following Arora et al. (2001) we call 'patent premium' (PP) the supplementary gain in case of a patent is taken (PP could be negative):

$$PP = PB - PN.$$

The firm will take a patent if and only if the net gain is positive. The net gain is the patent premium minus the cost of taking a patent.

As the patent mechanism consists in restricting competition, the patent premium is derived in two steps: it results from the impact of the patent (PAT) on the degree of competition (COMP), and from the impact of reduced competition on the market price:

$$PP = \frac{\partial \Pi}{\partial COMP} \cdot \frac{\partial COMP}{\partial PAT}$$

However patents do not exclude totally competitors while, even in the absence of a patent competition might not be perfect: the effect is not black and white. Patenting involves disclosure; which can trigger inventing around and follow up inventions by competitors; which then limit the control of the inventor over the market. Too much disclosure might even result in a negative patent premium. The easiness to imitate a non-patented invention, which is

---

Box 3.3. A sequential choice model of innovation and patenting

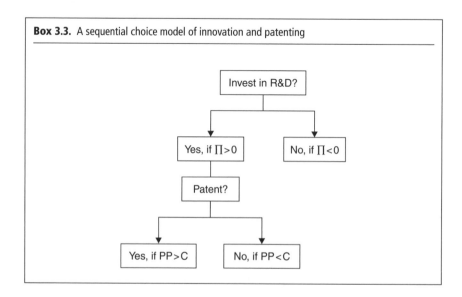

kept secret, depends on a range of factors, notably, how much of the invention is visible to competitors that the inventor would prefer to keep secret, and how much can be reverse engineered. The difficulty to 'decompile' an invention differs depending on the type of invention: products are less easy to keep secret than processes; technological complexity, which hampers imitation, is higher in aerospace industry that in automobile industry, etc. If the invention is easily reverse engineered the inventor will more likely patent it, except if he or she has other means of protection (e.g., brand name or manufacturing facilities that are difficult to emulate).

In turn, the impact of the market power of the patentee on profits depends on the price elasticity of demand. If demand is highly price elastic, for instance because there exists substitutable goods, then the seller cannot charge a significant premium. Consumers would then turn to goods rending similar services or to lower quality goods instead of paying the premium. Such a situation prevails in many mainstream markets such as textiles or commodities. The position of the patent holder is much stronger when demand is not price elastic, as it is often the case in higher quality goods for which customers highly value certain services provided by the technology (e.g., drugs). Low price elasticity is to be found also when a technology has few substitutes, such as an essential invention without which some products could not be manufactured. That happens notably for inventions which are components of technical standards, for instance in the telecoms or computer industries.

At the first stage of the model of sequential choice, the firm has to choose whether to invest in R&D or not. The decision depends on the impact of R&D on profits. The revenue generated by the invention should at least cover its cost, including R&D, so that the net profit is positive. In turn that depends on the marginal productivity of R&D (e.g., the likelihood of being successful, and the value of the invention). If the marginal productivity of R&D is high, a small change in R&D triggers a significant increase in expected profits.

Overall, the patent system could be considered as effective at the individual company level, if a significant proportion of inventions are pursued thanks to patents, which would not have been pursued without patents. That will happen in particular when the patent premium and the marginal productivity of research are high. Whereas the latter criterion applies more to high tech industries, with abundant technological opportunities, the former applies rather to traditional industries with an established patent tradition.

The way the patent system works is illustrated in Figure 3.1. Let's assume that the requested return for a project to be undertaken is equal to 8. All projects with an expected return lower than 8 will not be considered. If the company chooses in each case the strategy (patenting or not patenting) that maximizes the return, then project number 8 only will be dropped. The patent premium is positive in all but two cases (projects 2 and 6). What is the effect of patents on R&D? First, certain projects are insensitive to the possibility of patenting, as

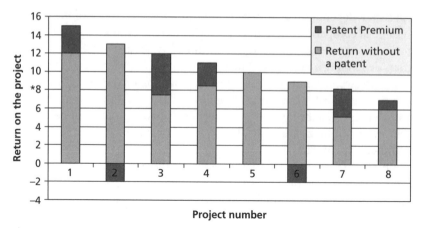

Figure 3.1. Effect of patents on the decision to undertake R&D
Note: Minimum warranted rate of return.

their return without a patent is sufficient (projects 1, 2, 4, 5, and 6) or their return is insufficient in any case (project 8). Certain projects like 2 and 6 will be implemented, but as the patent premium is negative the inventions will not be patented (taking a patent would effectively reduce the return of project 6 so that it would then not be implemented). For those projects which would have been undertaken even with no patent but which are patented, the effect of the patent is only to increase the profitability: patents have a windfall effect. The positive effect of patents on R&D is exemplified by projects 3 and 7, which are not profitable enough for being implemented without a patent but which pass the threshold if patented.

THE EFFECTIVENESS OF PATENTS

Empirical studies on the effectiveness of patents are not that many, and overall are not fully conclusive. They can be classified according to the research question they address:

1. To what extent and for what purpose do innovative firms use patents or other means of protection?
2. Does patenting add value to innovations?
3. Do patents induce further R&D and innovation?

Answers to the first question, the use of the patent system by industry, have been sought mainly through business surveys, asking essentially to respondents their use of and views about patents. The first such exercise, the so-called 'Yale survey' (Levin et al. 1987), conducted in 1983 in the US, generated some surprise when it was published, as it found that firms do not, in general, regard patents as very important to protecting their competitive advantage. Respondents ranked other means of appropriation, notably secrecy, lead time, reputation, sales, and services much higher than patents. These results were confirmed by a similar survey conducted in 1994 again in the US, the Carnegie Mellon Survey (CMS; see Cohen et al. 2000). That survey also showed the role of complementary manufacturing facilities and distribution services for securing rewards to innovation, as they endow the innovator with a certain degree of market power. Similar inquiries have been conducted in Europe as well since the 1990s, either as surveys addressing specifically patent issues (e.g., PACE in Europe; INSEE in France; and ISI-Fraunhofer in Germany), or as a special section in innovation surveys, notably the Community Innovation Survey (CIS) coordinated by Eurostat. While not dismissing the economic role of patents, these surveys show that patents should be considered as one component only in the appropriation strategy of firms, and often not the most important one. They also show that patents are used for a range of purposes that are not reflected in the simplest economic models. Major results from these surveys, which are similar across countries, are as follows:

- Patents are deemed effective for securing the return from inventions in certain industries only: chemicals, biotechnology, and drugs. In addition, they are effective for protecting new products (but not processes) in medical equipment. They are moderately effective in machinery, computers, radio/TV, auto parts, and, for process innovations only, in glass and petroleum. They are deemed not to be effective in other industries, including certain high tech sectors such as electronic components, instruments and aerospace (Cohen et al. 2000, consistent with other studies conducted in Europe). Accordingly, the share of product innovations which are patented in Europe is very high for pharmaceuticals (80 percent), chemicals, machinery, office and computer equipment, and precision instruments (50 percent), but very low for transport and telecom services, other transport equipment, basic metals, and textile/clothing (less than 20 percent) (Arundel and Kabla 1998a). The proportion of inventions which were patented in 2003 by firms of the OECD area having at least one patent in EPO (hence a rather biased sample) was about 66 percent for all industry, ranging from 50 percent in the audio-video-media area and 54 percent in the computer area (which both include software) to 72 percent in biotechnology and 73 percent in polymers (according to a survey conducted by the EPO with overall 481 respondents; see EPO 2005).

- Patents are more effective for product innovations than for process innovations. According to a survey covering the years 1990–93, large companies operating in Europe patented about 36 percent of their product innovations against 22 percent of their process innovation (Arundel and Kabla 1998a).

As processes are not as easily accessible to competitors as products, they could be kept secret more effectively without patents, and would suffer more from disclosure through patent documents.

- Patents are more often used for protecting radical innovations, based on R&D, than for protecting more marginal inventions based on other means (Licht and Zolz 1998). This is consistent with the inventive step required as a condition for obtaining a patent. This is also consistent with the fact that science-based inventions are more codified—they rely on known laws of nature—which make them easier to imitate than inventions associated with know-how and practical, uncodified knowledge.

- The major reasons for firms to patent are, according to the CMS survey conducted in the US (by decreasing order of importance): to prevent copying, to block competitors; and to gain freedom to operate (i.e., by not being blocked by others using threat or litigation based on their own patents). Gaining reputation and licensing revenues are less frequently cited. Similar ranking is given by surveys of European companies. The opinion firms have on the ability of patents to prevent imitation is the best predictor of the share of their inventions that they actually patent (Arundel and Kabla 1998b). It appears also that all these reasons are not independent from each other. Patents are part of a broader protection strategy with different components for various inventions or even the same invention. For instance, secrecy can prevail at early stages of research, then certain parts of the invention can be patented, while the underlying process is kept secret or protected de facto thanks to high entry cost (e.g., sunk cost of building manufacturing facilities, distribution networks, or reputation).

- Firms tend to patent more of their inventions when they are confronted with more intense competition. This is exemplified by the mobile telephone industry—see Box 3.4. Weaker competition (due, for example, to regulation or high entry cost) provides protection other than IPR to the innovations of incumbents, which then have little reason to incur the cost of filing IPR and disclose their technology. However, as patents in turn reduce ex post the degree of competition on a market, it is difficult to observe correlation between patenting and competition at the market equilibrium.

- Firms which export part of their production tend to patent more. Companies are usually exposed to more intense competition on foreign markets than on their domestic market, as they do not enjoy any home advantage, such as reputation or sales network, and exported goods are more subject to imitation in different countries.

- Large firms take more patents than small ones. In Germany in 1995, the share of innovating firms having patents was 10 percent among the small (ten to 49 employees) firms against 67 percent among the large (10,000 employees and more) firms (Licht and Zolz 1998). In France, only 15 percent of manufacturing firms with 20–49 employees had patents in 1997, against 80 percent of firms with more than 2000 employees (François and Lehoucq 1998). Could one infer that the patent system is biased towards large players? Not directly. The positive correlation with size holds for many activities, including R&D. That is partly at least a statistical artefact: the probability *ceteris paribus* that any event will happen in a given population is higher when the population is larger—and the same holds for a large firm as compared with a smaller one. The probability of having, say, an employee with red hair is higher in a company with 10,000 staff than in company with ten staff, and that has nothing to do with possible discrimination in recruitment. Hence, the fact that the proportion of companies having patents is lower among small companies than among large ones does not demonstrate, in itself, that the patent system is unfair with small companies.

Box 3.4. Increased patenting due to strengthened competition: the case of mobile telephones (based on Bekkers et al. 2002)

IPR used to matter very little in telecoms until the late 1970s. The industrial model was centred on national posts, telegraphs, and telecoms administration (PTTs), which were state-owned, with very little competition for manufacturing as, 'the PTTs procurement procedures usually favoured national suppliers, and national industrial policy determined which firms received long-term supply contracts' (Bekkers et al. 2002: 1144). National operators had their own research facilities. IPR was of little use as it would permit increase sales or mark up in such conditions. 'Manufacturers could be forced to license patents to other suppliers at no costs by national operators with multi-supplier policies' (2002: 1144). Much of the research expenditure of manufacturers was paid by the operator, leading to mixed ownership of the resulting IPR. Faced with no competition, operators had no incentive to protect their own IPR. This situation changed dramatically in the 1980s, due to the following factors:

- A liberalization trend worldwide led by the US (through privatization and deregulation).

- The increasing importance of technical standards for telecoms as the increasing demand for international communications required standards for interconnection, and the introduction of data communications between computer systems (not only voice) required interoperability.

- Standards were increasingly covered by IPR because of high R&D costs and worldwide, competitive market for standardized equipment; there has been an increased role of manufacturers (which are more competitive than operators) in defining the standards has strengthened.

With the openness of markets to competition in the late 1980s, operators tried first (1988) to arrange an agreement 'in which manufacturers were essentially forced to give up all their IPR for free world-wide licenses for essential patents' (2002: 1147). Motorola pushed hard to get licences, with high fees and strong conditions, on its patents. The European firms (Alcatel, Philips, Siemens, and Nokia) were not as aggressive initially, but eventually they followed suit. The Motorola leadership in this move is explained by its higher awareness of the value of patents, coming from the US experience (the large damages awarded to *Polaroid vs. Kodak* and Texas Instrument's successful legal battles in 1985/86). Patents are now at the core of the industrial model of mobile telephony, and licensing and cross-licensing are widespread.

The second question is: do patents add value to innovation—is the patent premium positive? Using data from the 1994 Carnegie Mellon Survey, Arora et al. 2003 find that:

1. For most innovations the patent premium would be negative, which is the reason why so many innovations are not-patented (the average patent premium for non-patented inventions is estimated at 50 percent, meaning that those inventions would lose half of their value if they were patented);

2. For those which are actually patented the patent premium is significant (it is between 180 and 240 percent depending on the industry, meaning that patents can double, on average, the value of an invention);

3. The patent premium has a skewed distribution, and it differs largely across industries: it is higher in drugs, biotech, medical instruments, machinery, computers, and industrial chemicals.

Duguet and Lelarge (2005) use a simultaneous equations framework explaining both innovation and patenting. They conduct econometric estimates using the CIS data collected in France. The authors find that the value of innovations is significantly increased by patenting in the case of product innovation, but there is no significant effect for process innovations.

Some studies have investigated the economic value of patents using survey questionnaires (asking directly patentees their own evaluation) or estimating with econometric regression the effect of patent related events on the stock market value of companies. A major result from the first strand of literature (see, e.g., Harhoff and Scherer 2000, using survey data from Germany) is that the value distribution of patents is highly skewed: most patents are worth very little, while a small number have very high value. Bloom and van Reenen (2002) and Hall et al. (2005) find that the grant of patents has a significant effect on the stock market value of companies only when it is weighted by indices such as the number of citations received from subsequent patents, which reflects the importance of the invention. A doubling of the citations weighted patent stock increases the value of a UK company by about 35 percent (Bloom and van Reenen 2002), and an extra citation increases the value of a US quoted company by 3 percent (Hall et al. 2005). However these studies do not really measure the patent premium, but instead the value of the patented invention itself. It is not clear whether the inventions would have a different value if they had not been patented.

These studies do not show whether patents have been successful in triggering further inventive efforts. This is our third question. This is a difficult question to answer as it would essentially require a counter-factual: comparing what would happen in reality, when a patent is filed, with what would happen in the absence of a patent. Several approaches have been used for tackling the issue. The first one asks companies directly what fraction of their innovation projects they would not have conducted in the absence of patents. Mansfield (1986) did such a survey in the US in 1981–83. He found this fraction to be relatively high for pharmaceuticals (60 percent) and chemicals (40 percent), and very low for other sectors (less than 10 percent for firms in electrical equipment, primary metals, instruments, office equipment, motor vehicles, and others).

A second approach is to use sophisticated econometric methods for controlling for all identified factors and relationships between them, so that the connection between patents and innovation could be as much as possible identified, and the two directions of causality separately measured. Arora et al. (2003) find that patents have a positive impact on R&D expenditure in most industries, especially drugs. Without patents, business R&D would fall by 25 to 35 percent overall in the US. An increase in the patent premium (associated, e.g., with a stronger patent regime) by 10 percent would generate an increase in business R&D by 6 percent on average. The effect would be higher in biotech and pharmaceuticals, lower in semiconductors and electronics (4–5 percent). Duguet and Lelarge (2005) obtain similar results for France for product innovation, whereas patents have no impact apparently on process innovation. Hall and Ziedonis (2001), analysing the semiconductor industry, do not find that stronger patent protection since the 1980s is driving the innovation effort of firms. Patenting in this industry was driven by patent portfolio races aimed either at ensuring access to technology and not being 'held-up' by rival patenting of the same technology, or at strengthening bargaining power when negotiating the access to other technology. As a result of this kind of zero-sum game, the patent over R&D ratio has surged in this industry while R&D itself did not benefit much.

In the same line of research, Shankerman (1998) estimates the value of the cash subsidy to R&D conferred by patent protection, called the 'equivalent subsidy rate' (ESR). It answers the question: how much subsidy should the government grant to the company to maintain its R&D at current level if there was not patent protection? Estimates by Shankerman on French data (based on a model expressing the value of the patent as a function of the duration of the payment of renewal fees) give an ESR of about 24 percent. Arora et al. (2003) conduct a similar exercise with American survey data and obtain an estimate of about 30 percent. Again, there is much variation across industries, with chemicals, pharmaceuticals and semiconductors featuring the highest.

A third approach is to compare patent regimes across countries or over time and correlate them with economic or innovation performance, in order to identify a possible causality. Such studies are faced with several methodological difficulties. The number of countries, hence of estimation points, is quite small. Many other factors than patents have influence on economic performance and innovation, so that identifying the specific effect of patents is not straightforward. Patent regimes are complex sets of legal and regulatory features, and differences between countries cannot easily be quantified and summarized. Countries are extremely heterogeneous in terms of economic status and the connection between patents and performance might not be the same, say in the Netherlands and in Zimbabwe. Notwithstanding these difficulties, quite a few studies have been conducted in this field, with the overall conclusion that stronger patent regimes tend to favour economic performance, although the relationship is not that simple. We just report on three such studies (others include Gould and Gruben (1996), Thompson and Rushing (1999), and have similar settings and results as the two presented below).

Kanwar and Evenson (2001) explore the effects of patent strength on technological performance. Technological performance is represented by R&D intensity (the ratio of R&D over GDP). Patent strength is represented by the 'Patent rights index' (PRI) designed by Ginarte and Park (1997). That index captures and quantify major features of patent regimes at the country level: membership to international treaties (e.g., the Paris convention), coverage (e.g., patentability of drugs, etc.), restrictions to patent rights (e.g., working requirements), enforcement (e.g., preliminary injunctions, etc.), and duration of patents. The index ranges from 0 (no patent protection) to 5 (maximum protection in view of the criteria above, reached only by the US in 2000). Other variables susceptible of influencing R&D intensity are introduced in the equation such as education, macro-economic conditions, interest rates, and degree of openness of the economy. The estimation is conducted on 29 countries, mainly rich ones from the OECD, over two periods of time (average 1981–85 and 1986–90). The PRI turns out to have a positive and significant effect on R&D intensity, and this effect is robust to changes in the specification and the estimation method.

Falvey et al. (2004) explain the growth of GDP by patent strength (represented also by the PRI). They have 80 countries over four periods of time (five years sub-periods between 1975 and 1994). They also control by broader economic variables and find an overall positive but weakly significant effect of patent strength on GDP growth. However, when controlled by the initial level of GDP per capita of each country in each sub-period, a more variegated picture emerges. Three categories of observations (countries by sub-periods) could be distinguished: low (less than US$670 in 1995), middle (between US$670 and US$10,829 in 1995) and high income (more than US$10,829 in 1995) countries. The impact of patent strength on GDP growth is positive and significant for the low and high income categories, while it is positive but not significant for the medium income category. In fact, the medium income category includes only a small number of observations (17 for a total of more than 300), and no country stayed in this category for all the period. It seems to be more of a transitory group, made of countries progressing from the low to the high income category. Accordingly, Falvey et al. interpret their results as follows. In early stages of development, the growth of countries requires imports of technology, that cannot be generated internally, through foreign direct investment and imports of goods. These are favoured by a secured patent regime. For high income countries, they generate internally much technology, and this innovative activity is encouraged by strong patent regimes.

Kang and Seo (2006) also test whether the strengthening of intellectual property rights stimulates innovation, as measured with the number of patents. They take into account complementary factors such as industrial structure, social capability, the stage of economic development and trade regime. Their quantitative analysis, which relies on a long-run panel data set of about 110 countries, reaches the following conclusions:

- IPR is positively and significantly related to innovation once other complementary factors are taken into account.
- The poorest countries are negatively affected by stronger IPR in terms of patent filings.

These results suggest that the effectiveness of IPR in fostering innovation varies across countries according to national economic and institutional contexts. For example, only countries with a per capita GDP above about US$9000 seem to gain, in terms of innovation, from the strengthening of IPR. That does not exclude, however, that they would gain in terms of productivity growth, due to an access to foreign technology facilitated by a stronger IPR regime.

Lerner (2002) studies significant changes in patent law in more than 70 countries over 150 years and correlate them with the number of patents granted in these countries (taken here as an indicator of innovation): He finds that strengthening patent rights have generated in general an increase in patent filings from foreign assignees, but had no effect on filings by nationals. That does not contradict the results above however, as foreign filings also reflect technology transfers, which are the dominant source of technology at early stages of development (where a majority of the 70 countries were at the time of observation).

The close association between stronger patent regimes and better economic performance that arises from data analysis is quite clear and robust. At early stages of development a strong patent regime encourages technology transfers, while at more advanced stages it fosters domestic inventions. In view of the relative degree of generality of the analysis it is based on and of the unavoidable lack of precision of such statistical studies, it should not lead however to the conclusion that it is *always* good to strengthen the patent system: There might also be an optimal level of protection, possibly different across countries and over time, beyond which negative effects start to dominate positive ones.

Overall, three robust conclusions arise from empirical studies, micro and macro. First, patents are quite effective in increasing R&D in certain industries, notably drugs and chemicals, less so in others. Second, patents are also taken for other reasons than simply avoiding to being copied, and it is to be expected that patents taken with these strategic objectives in mind are much less socially beneficial than others. Third, patent regimes contribute to economic growth and innovation: Through the import of foreign technology for less developed countries, through domestic inventions for more advanced countries.

3.4. **Inventions Disclosure and the Social Cost of Patents**

In the absence of legal protection for an invention, the inventor will often try to keep the invention secret. There is much evidence that secrecy is an important means of protection of inventions for many companies. It is one mission of patents to incentivise the disclosure of their knowledge by inventors, so that society would benefit more from it.

Business surveys confirm the widespread use of secrecy. A survey conducted in Germany with the reference year 2000 among 626 innovating firms (Husinger 2004) showed that 56 percent of respondents were using patents, 61 percent were using secrecy, and 41 percent were using both. As an independent use of the two means would lead to 34 percent (56 percent of 61 percent) instead of 41 percent using both instruments, that suggests a positive correlation between them: Among innovating firms, a patenting firm is more likely than a non-patenting one to use also secrecy. The same study shows that market success of product innovation is well correlated with patents but not with the use of secrecy, which tends to show that the most valuable product innovations are patented and that secrecy applies rather to process innovations or pre-market stage of product innovations. This is consistent with the view that companies take patents mainly when they cannot keep their invention secret any more. Respondents to the Yale survey (Levin et al. 1987) give 'too much disclosure' as the major reason for them not to patent some of their inventions. Overall, respondents to the Community Innovation Survey (CIS) tend not to use patents as a source of information (only 3 percent of European innovators responding to the 1994–96 CIS rated patents as an important source of information), but the proportion among most intensive innovators (who do formal R&D, cooperate with universities, and also file for patents) is much higher (Lelarge 2002). Responses to CIS also show that secrecy is preferred for protecting process innovations more than for protecting product innovations (Arundel 2001; Duguet and Lelarge 2005). This is the mirror image of the preference for patenting, and it makes sense as keeping a product secret is difficult, often impossible. Large car makers for instance have special departments whose mission is to reverse engineer new models issued by competitors, so that no innovation could remain secret. As noticed by a senior engineer of *Google*, 'A lot of our best ideas don't get filed as patents because patents eventually become public' (*Economist Technology Quarterly*, 17 March 2006, p.11).

The reaction of a particular company to a new EU environment legislation, the Registration, Evaluation, Authorisation and Restriction of Chemicals (REACH) illustrates the importance of secrecy, 'REACH is creating problems for companies such as Intel'. The legislation requires companies to document which chemicals they use in their production processes. But not everyone wants to reveal this information, for competitive reasons: 'This is raising a significant number of IP issues for us, because we use those chemicals in our manufacturing process', says an Intel spokeswoman. 'There are only a handful of people in Intel that have access to that information, and we don't want to register it publicly' (*Economist Technology Quarterly*, 12 March 2005, p.8). Patenting is a particular type of registration in that context.

Reading patents is the first step used by, for instance, Brazilian and Indian generic drugs companies when it comes to copying existing drugs: in order to understand the way the drug works; to find ways of manufacturing it. Further documentation (notably from the scientific literature) and experimentation are then necessary. Patents do not include the know-how, or further knowledge generated in the drug development process (after the patent application), but patents remain the starting point, a necessary step (Cassier and Correa 2005).

Secrecy is not the only alternative to patenting for a company that seeks to appropriate benefits from its inventions. For instance, much non-patented genetic material (e.g., genes, proteins) is stored in proprietary databases whose access is either prohibited to third parties or requires the payment of fees to the owner. That contrasts with patented genetic material, which is entirely disclosed in patent specifications hence becoming accessible to all potential users (although with restricted use).

It is one aim of the patent system to make disclosure a preferred option to the inventor. Disclosure facilitates follow up inventions (derived from the initial one), it facilitates the invention of substitutes to the initial invention, which increases the welfare of consumers and reduces market prices. As Werner von Siemens, a vocal advocate of patents in the late nineteenth century, once said: 'Because by patenting each new thought is carried into the world; one hundred minds will take it up, may steer it to completely different paths and make good use of it'.

The major source of new technology is existing knowledge, relating to science, technology, business or other. Economic modelling has integrated this idea since the mid-1980s at the macro-economic level with the theory of endogenous growth (Romer 1991) and at the micro-economic level with models of cumulative invention (Scotchmer 2004). New insights can be seen as essentially combinations of existing ideas (Weitzman 1998). Technical change is accelerated, and less costly, when researchers have access to more knowledge, to more ideas that they could combine, recombine and test. The effect of patents in that regards is twofold. On the one hand, disclosure makes available to the public, notably researchers, knowledge that would otherwise be kept secret. On the other hand, patents restrict the use that can be made of knowledge and might hamper research when they substitute to other forms of publications which are less restrictive in terms of authorized use.

That process can be represented in the form of a production function relating inputs to the output (new knowledge). New inventions (dA) result from the application of labour (R: The number of researchers) to existing knowledge (A: the number of existing inventions available to researchers, of which a fraction $q, 0 < q < 1$, is published) with the following linear process:

$$dA = s \cdot R \cdot q \cdot A,$$

where s is a productivity parameter, which reflects the average capacity of a researcher to draw new ideas from available ones. In that framework, the two effects of patents are captured: increased availability of knowledge results in increasing q, while increased difficulty to use available knowledge results in a reduction in s. The total effect of patents on the rate of new inventions dA depends on their relative impact on q and s.

There are limits to what research can be done with patented knowledge without the consent of the patent holder. Disclosure is not synonymous with public domain. The possibility of using patented inventions for doing further research without having to request the consent of the patentee is called the 'exemption for research use' (Dent et al. 2006). Its definition varies across countries. In most European countries, research aimed at testing the technology itself, in order to check the validity of the patent or to prepare negotiations for a possible licensing agreement does not necessitate the consent of the patent holder. For drugs also, in certain countries manufacturers of generics are allowed to test patented drugs with the view to improve on them some time prior to the expiration of the patent. This is all the more important as it takes years for bringing a drug from research to the market, including clinical testing and legal proceedings, then if research cannot start before the patent expires it would mean in fact that the exclusive right on drugs lasts longer than the patent legal term. That clause exists in US law (the Hatch–Waxman Act of 1984) as well as in most European countries. It is aimed also at balancing the fact that in these countries the patent term is longer for drugs than for other inventions.

Universities benefit from a legal right, or in certain countries de facto tolerance, for using patented knowledge without requesting licenses. This is based on the legal acceptance of the use of patents for non commercial purposes, which is recognised e.g. in the United Kingdom or Germany (it is more restrictive in France) (IPR Helpdesk 2005). However, the fact that increasingly universities take patents on some of their inventions, develop products next to the commercialization stage has put that clause under threat, notably in the United States, as certain patent holders see it as unfair treatment of de facto competitors. This argument is even more striking in the context of an increased funding of academic research by the business sector, which blurs further the distinction between profit and non-profit motivated research. A decision in 2003 by the Court of Appeal of the Federal Circuit (the *Madey v. Duke* case) made it clear that there is no such thing as a 'research exemption' in the US that would apply specifically to universities, even for the purpose of doing basic research. The Court argued that doing research, basic or applied, was part of the core business of a university, and that the university benefited in terms of funding as well as prestige from the research it did. Therefore, university interests and not only scientific curiosity were at stake in the research (See Nelson 2004).

Licensing is the traditional response to concerns arising from the increased difficulty of using knowledge when it is patented. Licensing contracts give under certain conditions, notably royalty payment, access to patented knowledge to other parties than the inventor. However, it has been argued recently that licensing does not work fully in certain fields, notably biotechnology. The so-called 'problem of the anti-commons' (Heller and Eisenberg 1998) is as follows. Licensing contracts are costly to set up, negotiate, and enforce. These transaction costs add to the payment included in the contract itself. If they are too high, they can reduce, even suppress, the benefits that would accrue to both parties from licensing deals. In certain technical areas, the number of pieces of knowledge needed for each particular research is high. In biotechnology, for instance, research on a particular disease can involve dozens of genes and other living material. If each of these pieces of knowledge is protected by patents held by different entities, the researcher will have to negotiate with many partners, hence increasing the transaction costs. This is the anti-commons effect, generated by a multiplication and a dispersion of rights that deter actions requiring them to be used in a coordinated way. The reality of this effect is being debated. A survey of more than 70 biotech professionals (e.g., researchers from the public and private sector, entrepreneurs, etc.) by Walsh et al. (2003) could not identify one case where research would have been cancelled due to that effect (see also Chapters 4 and 7).

In theory, in most countries, patented knowledge cannot be used freely by others than the patent holder with the view to improve on the patented product. That does not mean that disclosure is of no significance. In fact it is an important activity of patent departments in industrial companies to analyse patents filed by their competitors (Sueur 2004, on the chemical industry). That can give rise to new ideas, sufficiently different from the original one to circumvent the restriction in use ('invent around'). Reading patents gives a general sense of the evolution of technology, of the potential and limits of certain research directions. Companies can orient their research efforts accordingly, in particular avoiding entering into research that would duplicate patented inventions of their competitors. The reduction in duplication of research is one important expected outcome of the patent system, coming both from disclosure and from the exclusivity conferred by the patent (even if you replicate successfully your competitor's invention you cannot commercialize it as it is already patented). This effect is positive to society as inventing twice the same thing is worthless.

Patenting and secrecy are not necessarily incompatible and can even be complementary as mentioned above. One particular invention could be either patented or kept secret: but new products and processes are usually made of several inventions, some of which are patented while others are kept secret. One can patent the product while keeping the process secret. The quotation above of Intel illustrates that point, as that company will patent all its new micro-processors (and actually seek enforcement through litigation), while keeping secret key aspects of the processes that allow producing them in cost-effective ways. One limit of the patent system is that disclosure through patents often does not allow the implementation of the whole invention by third parties as key parts of the technology are kept secret.

DEADWEIGHT LOSS OF CUSTOMERS

Assessing the patent system from the point of view of society, instead of the individual patent holder, involves taking into account the effect that patents have on customers and on other companies.

The first effect is that customers benefit from new products and from the reduced cost of goods due to new processes. This benefit comes at a cost, however, a markup inflating the price of patented goods which results in a deadweight loss, represented in Figure 3.2. Thanks to a process invention, the marginal cost of producing some good drops from C to C'. The inventor takes a patent and charges a price P which is higher than the marginal cost. The fact that the vendor charges a higher price means that fewer customers will be served. For customers who are not ready to pay a price higher or equal to C', this cannot be considered as a social loss, from an efficiency point of view, as their consumption would imply a subsidy from society, just for producing the good. The picture is different for customers whose notional price is between C' and P. Serving these customers at their notional price would allow covering the marginal cost, hence requiring no subsidy, and would contribute to cover the fixed cost of research, although not as much as requested by the vendor. It would therefore be socially beneficial to serve them, but will not happen as the market price is too high. This is the dead-weight loss generated by patents.

This situation is not a purely abstract construct. It is exactly the situation which is debated in particular about access to drugs for poor customers, from developing countries. Once the research cost has been paid (by customers from

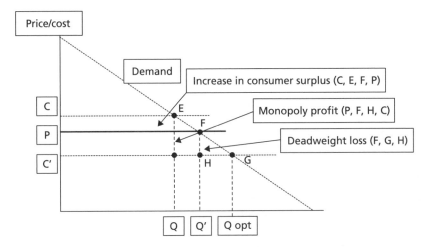

Figure 3.2. The effect of patents on static market equilibrium

Notes: P = price; C = Initial cost; C' = Cost after innovation; Q = Initial production; Q' = Production after innovation (function of P); Qopt = Optimal level of production after innovation.

rich countries), is it legitimate to deny access to the poorest customers even at the marginal cost?

A theoretical solution to that issue exists: price differentiation. If the firm could identify the willingness to pay of each individual customer and charge him or her accordingly, then it would find it beneficial to serve all customers whose willingness to pay is higher than the marginal cost C'. It is in fact an informational limitation that prevents companies from differentiating prices (i.e., discriminating among customers): it is not in the interest of customers to declare their willingness to pay, companies have only indirect and imperfect ways of assessing it. What they can do is identify certain observable characteristics of customers that could reveal their willingness to pay and thence make product offers in accordance. Then a second difficulty arises: customers with the highest willingness to pay could turn to those with the lowest willingness and offer to buy the good from them, hence setting up a second-hand market on which the original producer would have little control and which would drive down prices. The country of residence is an observable characteristic of a customer, it is correlated with revenue, hence with the willingness to pay, and it offers conditions that could limit the development of second-hand markets (with border controls). The latter property depends notably on the legal status of 'exhaustion of rights'. There is exhaustion of rights when the producer looses control over its goods after the first sale: national exhaustion means that the buyer has the right to re-sell the goods in the same country or in another country. International exhaustion means that the producer looses his or her rights in a given country after first sale in any country. International transactions involving re-sale of IP protected products are called 'parallel imports' (or 'grey trade'). On the basis of the Rome Treaty and in the name of the single European market, the European Court of Justice has ruled that IP rights are exhausted by first circulation of the product within the EU, but not outside the EU (or the EEA, which comprises the EU plus Switzerland, Norway, etc.). Hence, a second-hand market can develop within the EU, but not with other countries (Crampes et al. 2005).

Pharmaceutical companies are willing to segment markets, to differentiate their prices across countries as the market and regulatory conditions they face could vary (the price of prescription drugs is fixed by government in most European countries). A same drug can be several times more expensive in the US as it is in other countries, including its Canadian neighbour. That has triggered 'parallel imports' in the US, with many citizens just crossing the border to buy their medicines in Canadian drugstores or ordering on the Internet. There has been furious debate in the US between the pharmaceutical companies, claiming that such a practice is depriving them of the revenue necessary to fund their R&D, and opponents who see lower prices abroad as a proof that companies could well reduce their price in the US and still make profit. The Congress made it illegal in 1987 for anyone other than the manufacturer to import prescription drugs (Angell 2004). The risk of second-hand transactions booming and under-cutting legal sales in rich countries has been a major argument for drug companies to resist pressures for substantial price reductions to poor countries.

STRATEGIC PATENTING

The first effect of patents on companies other than the patent holder is to keep them out of a market. As competitors do not want to be excluded, they will take patents themselves in order to prevent that. Defensive patenting is aimed at creating rights that will not necessarily entitle their owner with a monopoly (excluding others), but will at least legalize its participation in the market, protect him or her against its competitors' patents. Hence, preserving the freedom to operate is another important reason for patenting, which is given a high ranking in business surveys; under the heading 'preventing suit', the CMS puts it third, immediately after preventing copy and blocking competitors (Cohen et al. 2000) which is in a way the symmetric move (that restricts on purpose the freedom of others to operate). This reason can be considered as strategic, in the sense that it is a response to competitors' behaviour rather than to the mere presence of an invention. As patenting constitutes one response to patenting by others, there emerges a strategic interaction which, like a repeated prisoners' dilemma, can typically display various equilibria. Let's assume that one company's motivation for patenting is to gain freedom to operate. Then if, in a given market, none of its competitors takes patents, this company will have no reason to do so itself. Once reached, this situation is self-enforcing, it is an equilibrium, if patents are costly to take and/or if retaliation is possible (i.e., 'if you restrict my freedom to operate I will restrict yours').

Conversely, if your competitors take out many patents a condition for your survival is to do the same. Hence the other polar equilibrium in that game is characterized by a proliferation of patents (there can be many other equilibria depending on specific assumptions regarding the game). If freedom to operate was the only reason for patenting, then the 'zero patents' equilibrium would be Pareto superior to the 'many patents' equilibrium, as the latter involves higher costs (notably the direct costs of working the patent system) while generating no more innovation.

That mechanism underlies the famous quotation by Fritz Machlup:

If one does not know whether a system 'as a whole' (in contrast to certain features of it) is good or bad, the safest 'policy conclusion' is to 'muddle through'—either with it, if one has long lived with it, or without it, if one has lived without it. If we did not have a patent system, it would be irresponsible, on the basis of our present knowledge of its economic consequences, to recommend instituting one. But since we have had a patent system for a long time, it would be irresponsible, on the basis of our present knowledge, to recommend abolishing it.

(Machlup 1958)

Such 'patent wars' occur in certain industries, like the chemical industry, in which certain segments are over-crowded with patents whereas others are patent free, ruled by secrecy and public domain—although it is the same firms in both cases. The semi-conductor industry is another prominent example in that regard (Hall and Ziedonis 2001). While few patenting was performed in that industry until the mid-1980s, certain companies (notably new entrants) started at that time to file many patents, triggering a symmetric response by others, so that after a few years systematic patenting was the strategy of the entire industry.

This view is confirmed by declarations from business officials. For instance, John Kelly, vice president of IBM, commenting on IBM's decision of releasing 500 patents in the public domain in January 2005, compares patent portfolios to missiles, and the patent race to an arm race: 'The layer of technology that is open is going to steadily increase, but in going through this transition we're not going to be crazy,' Kelly said. 'This is like disarmament. You're not going to give away all your missiles as a first step' (*International Herald Tribune*, 11 April 2005, p.11). Or Kab-tae Han, manager of IP at Samsung digital media's business, says:

Most Korean companies, including Samsung, did not pay much attention to IP before 1985. After that, major Korean companies realized that filing patents is so important after a number of them were involved in patent litigation in the US. We were asked by our top management to enhance our patent portfolio in a short space of time to cope with the situation.

(*Managing IP*, November 2004, p.viii)

That reaction, building weapons as a response to others' aggression, makes perfect sense from a private point of view: but from the point of view of society it looks like a waste of resources. However, even in that case, patents induce disclosure and generate some benefits that would not occur otherwise.

Nascent conflicts between incumbent firms resulting from the accumulation of such rights are often resolved in cross-licensing agreements, hence avoiding too much litigation. However, that does not eliminate the cost of setting up and maintaining huge patent portfolios. That has the effect also of increasing the cost of entry to the market to newcomers, which must accumulate a large patent portfolio in order to be able to resist to incumbents and avoid litigation for patent infringement.

DISTORTIONS IN PROFITS AND INVESTMENT

It is one purpose of the patent system to allow the emergence of rents on the market. As the effectiveness of patent protection differs across many dimensions, notably industry, the size of the patent premium will vary accordingly. It is higher for instance in the drugs industry than in the textile industry (Walsh et al. 2003). Hence, the patent system affects the distribution of profits across industries, and probably affects the allocation of investment accordingly. An economy with a stronger patent system would be expected to specialize more in drugs, as an economy with a weaker patent system would specialize, say, in textiles. This assumption has been tested by Moser (2005) who, using data from the mid-nineteenth century industrial exhibitions, shows that the suppression of the patent system in the Netherlands from 1869 to 1911 led to re-orientation of industrial research towards areas—mechanics notably— where secrecy and technical complexity offer protection.

Strong patent protection in certain industries might result in a high payoff, a private return which exceeds the social return, which could attract more investment in R&D than it is socially efficient. This is known as the 'common pool' problem (Loury 1979; Wright 1983), which is a special form of the general problem of rent dissipation (whenever a rent is created, rent seeking activities will tend to exhaust it). The mechanism is formalized in a patent race model.

Let us assume a (still to come) invention which would provide a rent R to the inventor. The cost of doing the research is the same for all participants to the race, r. Ex ante, a participant does not know whether he or she will win or not, hence the expected gross gain is:

$$EGG = p \cdot R,$$

where p is the individual participant's probability to win. The expected net gain for each individual participant is the difference between the gross gain and the cost:

$$EGG = p \cdot R - r.$$

We assume the probability to win to be the same for all participants (hence $p = 1/n$, where n is the number of participants). New participants will enter the race until the net gain is zero: $0 = R/n - r$. Hence the number of participants at equilibrium will be $n = R/r$. In this simple model, it is clear that each new entrant beyond the first one just adds cost to society (equal to r), without adding value: one research project would be enough for the invention to be done. An amount $R-r$ of resources is wasted in research, and a rent r would have been sufficient for attracting one participant. Hence a weaker patent would be preferable.

A similar mechanism is at work in patent races where each participant has a probability to invent first, hence win the race, which is an increasing function of his research expenditure. When fixing its level of research expenditure, each company takes into account the *positive* impact of its marginal Euro on its own probability to win. It does not take into account its *negative* impact on other participants' probability to win. Only the private gain, not the social cost, is taken into account in private decisions. Over-investment in specific fields of research can be favoured by the strength of patents in these areas.

Over-investment in research due to the extra-reward generated by patents is not easy to detect empirically. There have been well documented waves of over-investment in certain technologies, such as HD-TV in Europe in the late 1980s, in biotechnology in the US in the early 1980s or in Internet technology in the late 1990s. However, they could be attributed mainly to over-optimistic expectations (by government or by markets) regarding the immediate potential of the technology more than to a patent effect. Certain observers claim that such an effect occurs in the pharmaceutical industry, with the practice of 'me too' drugs—drugs having a similar effect as an existing one, operating with a similar mechanism, developed by companies following the demonstration of technical feasibility and economic viability made by a competitor. 'For instance, we now have six statins (Mevacor, Lipitor, Zocor, Pravachol, Lescol, and Crestor) on the market to lower cholesterol, all variants of the first' (Angell 2004). This is denied, however, by drug companies, which claim that each of these drugs has a differentiated effect or are effective in particular populations of patients. The counterpart of focusing research on excessively rewarding areas is that less rewarding ones are abandoned: hence the issue with orphan diseases, which the patent system has certainly not created, but has not helped to solve.

It is not clear, however, that duplication is really a problem, that all duplication of research is total waste of resources, or that patents reinforce the problem of duplication. Five arguments can be raised in this respect. First, parallel research might result in competing patented goods (enough differentiated for getting each a patent), generating at the end more competition on the market for goods, that will benefit to customers in terms of lower prices. Second, duplication might be partial only, resulting in slightly differentiated goods that bring diversity on the market (e.g., drugs having different patterns of effectiveness or allergenic effects depending on the characteristics of the patient). Third, there might be different ways of doing a given research, with none being ex ante clearly superior to others. Competitors will then pursue different routes, and the expectation that the invention will be made at the end is higher thanks to this diversity of strategies. Fourth, patent races trigger an acceleration of research, resulting in earlier inventions, and such gain in time might generate significant social gains. A case in point is decoding of the human genome in the late 1990s. There was a race between a public consortium, the Human Genome Project (aiming at the public domain) and a private company, Celera (aiming at patents). As a result of that race, extremely efficient techniques of automated sequencing of the genome were developed, which accelerated the process by several years. As much downstream research (determining the function of the genes and the relationship with various diseases) was dependant on that initial effort, the social gain can be considered as huge. Fifth, patent disclose inventions and could prevent duplicative research occurring because of the ignorance of companies about each other's research activities.

3.5. **Summary**

This chapter has investigated the rationale of patents and their economic role:

- The 'natural rights' theory, highly influential in legal circles, sees patents as a natural and fair reward to inventors. At the opposite, the utilitarian theory considers patents as an incentive, a policy instrument used by society for encouraging inventions. Patents are a response to the public good nature of knowledge, which makes imitation easier than invention.

- Patents are only one of the instruments of innovation policy, along with grants, prizes, subsidies, universities, and public laboratories. As compared with other instruments, patents (1) are more market friendly, leaving all technical and economic choices to firms and customers; and (2) restrict the diffusion of knowledge.

- Empirical studies support the view that patents are quite effective in encouraging innovation in certain industries like pharmaceuticals or chemicals, less so electronics, and have little direct effect in most other industries. A strong patent system favours innovation and growth, in poor countries as it helps the import of technology and in advanced countries as it encourages domestic innovation.

- Patents generate a 'deadweight loss', as customers willing to pay the marginal cost but not the mark up are excluded. Patents also restrict the diffusion of knowledge as they allow inventors to require fees to users. On the other hand patents disclose inventions which could otherwise be kept secret.

4 Patent as a Market Instrument

Dominique Guellec, Bruno van Pottelsberghe, and Nicolas van Zeebroeck

The previous chapter shows how patent policy, its design, and its relationships with other policy tools might stimulate innovative efforts. It mainly focuses on the economic and policy side of patent policies. In this chapter we analyse the market side of patents. We first investigate how and why users rely on patenting. The first section describes the various determinants of patent behaviour, ranging from traditional characteristics of firms (their size and market power) to the influence of their innovation and intellectual property (IP) strategies. The second section focuses on one particular reason underlying the recent surge in patenting: the development of a market for technology, as witnessed by the surge in licensing deals. The third section analyses the potential economic effect of licensing practices, and how competition policy may hamper socially detrimental practices. The last section reviews the various methodologies and tools that are available to assess the value of patents and exploit them financially. First it describes the valuation methods for patents, then it looks at recent trends in patent valuation and monetization, and finally it briefly summarizes empirical findings on the determinants of patent value.

4.1. The Various Uses of Patents

By endowing innovative companies with a temporary monopoly on a specific market, patents provide them with the opportunity to build an advantage over their competitors. There are various ways this market advantage can be used, beyond the direct extraction of a rent from customers. Patents provide a bargaining power to their holder, which can be used for various purposes:

- *to restrain the power of suppliers* by owning key technology elements in another part of the value or technological chain—a strategy used for instance by Nokia with loudspeakers (Reitzig 2004a);

- *to 'freeze' a technology*: patents can be used to prevent the development of to a particular market or technology. This strategy refers to 'technology suppression', which consists in freezing a technique by obtaining and controlling it (exclusively) and keeping it out of the market to avoid cannibalizing the firm's currently exploited technology (see Box 4.1);
- *to guarantee (to the holder) freedom to invent*: in this case patents are applied on future technologies, with market potentials but a high uncertainty. It allows firms to secure in advance the possibility to exploit an invention in the future;
- to build *negotiating power* for the acquisition of complementary inventions, or for an easier access to the market;
- *to avoid being 'invented around'*: the company builds a thicket of patents around a key patented invention, designed to block competitors;
- *to secure freedom to operate and avoid potential litigation*: as an increased number of companies are litigated for infringement on their own unpatented technologies, but being patented by a third firm (a competitor or a troll);

BOX 4.1. Einstein's refrigerator

The Einstein refrigerator is a type of refrigerator co-invented in 1926 by Albert Einstein and former student Leó Szilárd, who were awarded US patent number 1,781,541 on 11 November 1930 (the patent number 282,428 entitled, 'Improvements Relating to Refrigerating Apparatus' was applied on 16 December 1927 in the United Kingdom and was granted on 5 November 1928; and the priority application was from Germany on 16 December 1926). The machine is a single-pressure absorption refrigerator, similar in design to the gas absorption refrigerator. The refrigeration cycle uses ammonia (pressure-equalizing fluid), butane (refrigerant), and water (absorbing fluid). The Einstein refrigerator is portable, made of inexpensive, non-moving parts, operates silently, and is very reliable. However, leaks of the toxic ammonia caused problems among the earlier models.

Einstein undertook this invention as a way of helping along his former student. He leveraged the knowledge he had acquired during his years at the Swiss Patent Office to get reliable patents for the invention in several countries. But the refrigerator never came into commercial production. The new refrigeration technology was eventually 'cooled off' by the Depression and the invention in 1930 of Freon (CFC), which fixed the vapour-compression process as the refrigeration standard.

However, despite this apparent failure, two factors seem to suggest that the invention potentially had a substantial value.

First, the most promising patents were quickly bought by the Swedish company AB Electrolux to protect its refrigeration technology from competition. Hence, the inventors earned some (undisclosed) financial rewards for their patented invention.

Second, the underlying concepts and technical specificities became the basis for cooling nuclear breeder reactors: quite an important social return.

(Silverman 2001; Wikipedia at: http://en.wikipedia.org/)

- *to set up 'picket fences'*: through which a company directly reacts to a new patent filed by another company;
- *to prevent others from acquiring IP rights*: companies file defensive applications or publications that they let fall into the public domain to avoid potential appropriation by competitors; and
- *to 'create a smoke screen'*: filing applications on technologies that will not be exploited by the company in order to disturb competitors into thinking that they might.

All these tactics induced by strategic reactions to competitors' behaviour generate more patent applications. The influence of market forces is well illustrated by the semiconductor industry where the enormous financial consequences of the potential injunction of a pending patent litigation to shut down a manufacturing plant—should it be even for a very short period of time—urge industry players to file patents which are necessary as a dissuasive bargaining instrument. Similarly, patents are sometimes necessary as an exchange currency to access a specific third-party technology, which can only be traded against other IP rights.

From a firm's perspective, facing a new and promising patent obtained by one of its competitors, there are not so many ways of avoiding the loss of competitive advantage and market shares. The first possibility would naturally be to get the patent revoked through legal procedures such as opposition (at least in the European Patent Convention framework) or litigation. Should such initiatives fail or appear too hazardous and costly, the only other options are to buy or license in the new patented technology or to invent around the patent. The 'invent around' strategy (Glazier 2000), consists in developing a product or process that does not infringe on the competitor's patent but still allows the firm to enter the same market. The resulting invention is then patented in itself.

The fear of inventing around strategies invites initial patent holders to apply for patents with the broadest scope possible—i.e., making them difficult to invent around. Otherwise the inventor will eventually have to invent around invented-around patents in return. The former strategy implies to draft any patent in such a way as to block the avenue to as many alternative inventions as possible or to invent around your own patents and file patents covering the resulting incremental inventions. Together, these behaviours induce a snowball effect where competitors keep on patenting for minor inventive steps, either around each other's patents, or around their own patents.

Alternatively, the 'picket fence' strategy consists in reacting to a new patent granted to a competitor by protecting any incremental improvement representing all preferred utilizations of the patent, making it simply impossible to exploit as such by its initial holder. The picket fence puts thus the competitor in a situation to force cross-licensing deals and obtain therefore access to the core technology.

One way of avoiding such reactions from competitors to a new invention would actually be to protect it without disclosing it. This is exactly what 'submarine patents' aim at doing: to circumvent the disclosure obligation resulting from a patent application and spare time in the patenting process for the market to mature. This strategy consists in filing a patent application and keeping it pending by using any possible procedural delay and filing a string of amendments to the claims as long as the most promising applications and markets based on the invention are not clearly identified. As soon as they are, the patenting process is allowed to go to its end and the patent gets granted. Once the patent is granted, its holder may turn back to its competitors, and ask for royalties back from the date of application (see Chapter 6 for more details on patent drafting strategies).

Although submarine patents have been restrained by the TRIPs signed in 1994, which limited the life of patents to 20 years from the date of application and made pending applications public after 18 months (which was already the case in Europe), they are still a reality. As patent applications are published, submarine patents take the form of very large applications with hundreds of pages and/or claims, making the actual invention virtually invisible and almost unsearchable as it is hidden in the middle of irrelevant information. Once the market has matured, only the relevant claims are kept from the initial application and can get granted very fast, becoming visible to all.

An opposite—and extremely defensive—strategy consists in filing an application and withdrawing it before it gets granted. Firms consider such a strategy when they are willing to prevent any third party from filing a patent on their invention but are not ready to incur the costs of maintaining and enforcing a patent.

As a matter of fact, inventors do not always file patent applications with the intention to get their application granted. In addition to the defensive reasons evoked above some firms file applications—sometimes many—with the intention of polluting a technological field by creating uncertainty. In extreme cases, such strategies can go as far as pure patent flooding, or filing excessively large applications that become part of the existing art once published and so generate a smoke screen in the field.

4.2. **Licensing and the Market for Technology**

It has been argued since Arrow (1962) that innovation is subject to a particular information asymmetry that creates moral hazard and makes it extremely difficult to trade. Trading implies that the buyer has a fair idea about the technical characteristics and the market potentials, so that he is able to assess the value and estimate a price he would be ready to pay. In view of possible distortions of information by the seller, and of the fact that the potential value of an invention is dependant on its competitive and technological context, it can hardly be sold 'off the shelves', just from a catalogue. The potential buyers of a new technology might require more information before realizing the purchase, but once they have had access to the invention they may decide not to buy it but to exploit it for free instead, especially if the invention was protected by secrecy only. Knowing that and in order to prevent it, the owner of the invention will not allow access to the technology to other parties, hence making any deal impossible. Private contracts agreed ex ante for protecting secrecy, including non-disclosure agreements, are an imperfect remedy as they are costly to enforce. If the potential buyers choose to use the technology they tested without entering into licensing, the inventor has to go to court and to demonstrate that he or she had the invention before (without public documentation as it was kept secret).

Patents offer a response to overcome this difficulty. They include a clear description of the protected invention along with a certification of the legitimacy of its owner, that could be submitted to courts in case of alleged illegal use of the invention by another party. Non-patented components of the technology, or 'know-how' (or tacit knowledge) are often included into licensing agreements having patents as a backbone. By strengthening the position of inventors, patents make them more willing to enter into such transactions. It is a major advantage of patents over other means of appropriation of technology: they enable the development of market transactions of inventions (see Arora et al. 2001).

A licence is a contract by which the patent holder authorizes another party to use its patent under certain conditions (see OECD (2005) for a general presentation of licensing practices). Licensing contracts are extremely variegated. The use of the invention can be restricted to certain markets defined, e.g., in terms of geography or in terms of goods. Licenses can be exclusive (one licensee only for the technology and for the world), geographically exclusive (one exclusive licence per geographical area) or non-exclusive (several licensees per geographical area). The most frequent compensation to the licensor is the payment of fees by the licensee, but the compensation can also be a licence on another patent granted by the licensee to the licensor (cross-licensing). The contract may allow the licensee to modify the technology or not, possibly with grant back provisions).

Licensing contracts are complex (Mervin and Warner 1996). The most significant factors that are taken into account are the extent of the rights to use the technology, the term of the agreement (or duration), the geographical scope, the degree of exclusivity, the allowance to sub-licence or not, the termination clauses, the inclusion of trademarks or trade names, the inclusion of technical assistance, the payment terms, and relational issues, such as the waiver by the licensor of all claims it may have against the licensee for prior use of the products.

LICENSING-IN AND LICENSING-OUT

Why would patent holders be willing to license their technology to other parties? The following (non-exclusive) reasons justify the 'licensing-out' of a patented technology:

- *Obtain licensing revenues*: this is the major motive for companies specializing in research who do not themselves implement their inventions. Overall, however, that objective is not ranked high by business surveys, as witnessed in the US (e.g., CMS or Carnegie Mellon Survey) and Europe (e.g., CIS, the Community Innovation Survey).

- *Reduce production costs*: a company with production facilities might find it cost-effective to have its invention put into production by other companies which are more efficient in manufacturing, simply because they already have the required complementary assets to produce and commercialize the invention. This happens if the premium that the licensors can charge is higher than the profit margin they would have by implementing the technology themselves.

- *Reduce risks for customers*: licensing-out to competitors provides a credible commitment by an innovator to abstain from exploiting its monopolistic power and hence inducing customers to adopt its own technology (Farell and Gallini 1988). The commitment may ensure price stability, minimum quality threshold (Shepard 1987), and stability of supply.

- *Gain access to other markets*: either for regulatory reasons (trade barriers) or due to transportation costs, some national markets are of restricted access to foreign manufacturers. In that case, innovators can license their technology to local producers as a way to enter, indirectly, the market.

- *Expand the range of products related to the technology*: inventions made in particular industry could have applications in other industries. As the inventors have no manufacturing or marketing competencies in these other industries, they might prefer to license the invention rather than entering directly the market. This reason could particularly apply to small firms, which have a narrower range of products and might have more difficulties integrating into their product line an invention which is out of their core business.

- *Gain access to other companies' technology*: licensing its own technology might prove useful if a company wants to access a third company's technology. This bilateral exchange would induce a mutual access to each company's technology, especially when the two technologies complement each other. This is a case for cross-licensing.

- *Standard setting*: out-licensing can be a way to make a particular invention part of an industry standard, hence boosting its value (Grindley 1995).

- *Compulsory licensing*: a company can be forced to license out its technology by courts or governments, either on competitive grounds or for political reasons (national emergency related to health or security, when the patent holders require too high a price or if they cannot provide the required quantity of product). Compulsory licensing is in fact extremely rare.

Motivations for the 'licensing-in' of a new technology largely mirror those for the licensing-out:

- *Save on R&D expenditures*: by using an existing technology instead of inventing their own, companies can avoid the costs and risks generated by R&D activities. In that case licensing-in substitutes to internal R&D, and actually avoids the 're-invent the wheel' syndrome.

- *Accelerate the innovation process*: by acquiring some existing technology instead of setting up in-house R&D projects, companies can save time to achieve results.

- *Expand the product range*: gain access to the technology needed for new products or processes and which cannot be developed internally. In that case, licensing-in complements internal R&D.

- *Avoid litigation*: if companies feels threatened by a patent holder who could litigate them for infringement, they might definitely prefer to license-in preventively.

- *Gain access to a standard* protected by patents.

Licensing payments usually have two components, a fixed fee and a variable component, a royalty over sales (in percentages) or units (fixed amount per unit). Usually, in such contracts, these two components are designed to share the risk between the parties and maximize their commitment and effort to the contract. The fixed fee reduces the risk taken by the licensor (who will obtain the revenue whatever the sales) and gives more incentives to the licensee to increase sales (in order to break even), while the variable component puts some of the risk on the licensor (so that its interest is to maximize the quality of the technology).

There are alternative methods for calculation of royalty rates, which may be preferred according to the context (Sullivan 1996). These methods include the 25 percent rule (of a licensee's pre-tax gross profit, which corresponds to a royalty rate of about 5 percent of the licensee's selling price), the market or industry norm approach (2 to 8 percent of net profit), the return on sales, the return on R&D costs, the income approach and the lost profits or reasonable royalty, which includes lost profits resulting from infringement, and/or a reasonable royalty for the use of the patented invention.

As licences are private contracts, there is no obligation of disclosure of the terms and companies are usually reluctant to publicise such information, considered as commercial and strategic information. Hence relatively little is known regarding the size and major actual characteristics of licensing markets.

Using various sources and making a range of assumptions, Arora et al. (2001) estimate the market for technology to be about US$25–35 billions in the mid-1990s, while the world as a whole would be about US$10–15 billion higher. This includes, however, more than patent licences, notably licences involving know-how and other collaborations (e.g., production and marketing). If one assumes that the share of licensed inventions have not changed since then (a conservative assumption), then the licensing market should have evolved roughly in line with business R&D expenditure. As business performed R&D has increased in nominal terms by about 50 percent between the mid-1990s and the mid-2000s, it would put the licensing market around US$40–50 billion in the US and US$15–25 billion in other regions in the mid-2000s. As much qualitative evidence points to an increase in the propensity to license technology during the 1990s, this estimate could well be somewhat conservative for the mid-2000s. According to Athreye and Cantwell (2005), royalty and licensing fees worldwide have surged from around US$10 billion in 1980 to over US$90 billion in 2003 (including revenues from licensed patents but also from copyrights and other types of intellectual assets).

According to a survey of about 380 respondents conducted by the EPO (EPO 2005c) with reference year 2003, an average of 11 percent of patents held by companies from OECD countries were subject to licences, the proportion being higher in the US (15 percent) than in Japan (8 percent) or the EU (11 percent). A study by Gambardella (2005) shows that within Europe, the UK has by far the most active market for licensing.

Licensing displays also clear variations across industries. It is much more developed in the information technology (IT), biotechnology, and chemical industries, with a focus of IT on cross-licensing. That could partly explain the edge the US has on Europe, and that the UK has on the rest of Europe, as all three industries are relatively more developed in the US, and biotechnology is more developed in the UK than in the rest of Europe (Gambardella 2005).

Small firms and universities are more likely to license their patents than the larger ones (Gambardella 2005). First, because patents with a higher scientific content are more subject to licensing than others and, second, because universities and small high tech firms usually do not have the required complementary assets or focus to introduce new products or services onto the market (manufacturing, marketing). Only 15 to 20 percent of the patents applied by universities or small firms are licensed, however, against less than 5 percent in medium or large firms.

THE BENEFITS AND COSTS OF LICENSING TO SOCIETY

The effects of licences on the economy are variegated. They first induce a more effective diffusion of technology. The invention, representing presumably the most recent and efficient technology, is made available to more firms than when it is only exploited in-house by the inventor. Second, licences are at the root of improved productive efficiency, as they lead to a more intense vertical specialization—a split between research and manufacturing. Some companies are indeed better at inventing and performing research, while others are better at manufacturing activities. In the absence of markets for technology, the two activities are not separated, so that many companies have to conduct activities for which they have no specific advantage.

The growth of markets for technology has actually coincided in several industries with the emergence of firms specialized in research. This was the case in the chemical industry (with the specialised engineering firms (SEF)), in the semiconductors industry with the 'Fabless' companies, or in the biotech industry (which would not exist as a separate industry from the pharmaceutical industry if no patents and licenses were available (see Arora et al. 2001). Group Lotus PLC is another example of vertical specialization, in the automotive industry (Bosworth and Pitkethly, forthcoming). The main activity of the Lotus Group is to perform research and develop new techniques for selling engineering projects to their clients. The second activity of the Lotus Group is to license its technology.

The development of markets for technology should in turn increase the demand for patents. Existing companies will patent more of their inventions, with the prospect to licensing them out, rather than attempting in-house exploitation. New companies specialized in research are flourishing, and they have a much higher propensity to patent their inventions than large, vertically integrated firms, who exploit themselves their technology and are more inclined towards secrecy. All in all, more knowledge is publicized than in the vertical integration model, which in turn fosters follow-up inventions. In the same vein, licensing is one possible solution to the 'anti-commons' effect, as it facilitates access to technology.

Non-exclusive licensing can foster competition as more suppliers are on the market, to the benefit of customers. In that case, licensing reduces the mark-up of sellers as compared with the non-licensing (monopolistic) situation (there is a rent dissipation effect): on the other hand, the volume of sales is higher (the revenue effect). The patent holders, who maximize their total revenue, will favour non-exclusive licensing only if the latter effect is higher than the former one.

In certain cases, however, licensing out might reduce the level of competition. It happens when the availability of one technology for licensing deters research to be undertaken on alternative technologies that would be then marketed by competitors. Hence licensing is used to build an upstream, indirect, monopoly position which can be used by its holder to collect a higher mark up than under competitive conditions. It might also happen in the frame of cross-licensing deals, by which two or more companies controlling most of the technology needed to enter a particular market decide to go hand in hand, therefore possibly raising barriers to possible new entrants. Competition authorities are particularly aware of these possible problems, which is the reason why they tend to monitor licensing deals quite closely.

4.3. **Patents and Competition Policy**

Schumpeter argued that innovation is connected to the degree of market power. A monopolistic position provides market power and the extra revenue that helps funding innovation. Since Schumpeter the economic literature has been ambivalent regarding the connection between innovation and competition. The current thinking can be summarized as follows (Encaoua and Ulph 2000): innovators are motivated by the perspective of high expected profits, which are higher with a monopolistic position. On the other hand, companies in a situation of solid monopolistic position (either natural monopolies or government enforced monopolies) tend not to be motivated to innovate as that would not significantly increase their already substantial profit margin. The threat, or reality, of competition forces incumbents to innovate, while for newcomers innovation is their ticket to enter the market.

In other words, innovation tends to destroy competition, while competition spurs innovation. In a dynamic setting, notably in fast moving technical areas, the monopoly position possibly provided by successful innovation is temporary only as new inventions come fast, with superior technology taking over the market, leapfrogging incumbents. In this cycle, patents play the role of strengthening the market power that accrues to the successful inventor, hence reinforcing the incentive to innovate ex ante, but possibly weakening the incentive to innovate for the winner, at least ex post.

A patent is a right to exclude competitors. On face value, it therefore has direct anti-competitive effects. A product will have a higher price if it embodies a patented technology, due to the market power conferred by a patent. Although this is the rule of the game, it is not the end of it, as competition, notably induced by follow-up inventions, is a dynamic process, which should be seen with some time perspective:

- A patent rewards an invention, a new technology, which sometimes results in creating a new market. In that case, the effect of patents from period t to $t+1$ is not to restrict competition on markets already existing in t, but to create a new market—possibly temporarily monopolized, but still better than no market at all.

- Patents offer a substitute to secrecy and involve disclosure, hence they encourage further innovation—i.e., competition of new products against existing ones.

- Patents can serve the creation of new companies by protecting them from competitive weapons given by incumbency (like size, brand, or sunk costs).

In addition, in the context of knowledge-intensive industries, competition takes specific forms, resulting in a different role assigned to patents (see Encaoua and Hollander 2004). Competition is less based on prices and current market shares (i.e., a static perspective) and more based on new products and techniques and tomorrow's market shares (i.e., a dynamic perspective). The market power criterion is more fragile, or 'contestable', as the state of play can be reshuffled overnight by new technologies. Substitute products are not the current competitors but the ones to be put on market in the near future.

In these industries also, standards are more important, as products are made of many independent components or must be used with other, complementary products that have to be compatible. Being the standard setter, or at least influencing the standard setting process is key to companies. The need for speed and compatibility encourages companies to set up alliances with each other as part of the competition process. Overall, there is more competition *for* markets than competition *on* markets. In that context, patents are not just a means to exclude competitors, they are an instrument used by incumbent firms for raising entry barriers, or alternatively used by new entrants for penetrating markets, they are used in standard setting processes and for making alliances. This diversified and dynamic role of patents renders their effect on competition more complex.

Patents can be used in anticompetitive strategies, whose aim is to exclude other companies (competitors) from the market. From a legal point of view, 'abuse' in that context could be defined as patentees seeking exclusion rights beyond what they are entitled by their patent and by patent law. From the economic point of view the boundary between 'fair use' and 'abuse' of patents is not always clear cut. A patent is a right to exclude: beyond which point is this right 'excessively implemented'? In a utilitarian perspective one could characterize this point according to the expected impact of the patent on competition and innovation. The right is 'excessively implemented' if its effects on prices and on innovation are detrimental, for example if the increase in prices is not balanced by a commensurate in innovation or, more obviously, if one sees both a supplement in price and a slowdown in innovation due to the behaviour of the patent holder. Of course there is no clear-cut generally applicable rule in that matter. Certain conducts of patent holders are susceptible of hampering competition and innovation, whereas others seem to be good for society at large. Here are examples of conducts with possible negative effects.

Patent thickets

These are 'dense webs of overlapping intellectual property rights that a company must hack its way through in order to actually commercialise new technology' (Shapiro 2001). A patent thicket can be built intentionally by a single firm, or can result from competition between innovators seeking to maximize their respective protected areas.

Prima facie, a patent thicket is not against the law if each individual patent is legal. There is, however, some evidence that in certain cases such large numbers of patents are being used just as barriers to entry: Large numbers of overlapping patents make it extremely difficult to assess the scope of rights that their holder is entitled to. Hence patent thickets are used for extending the exclusive rights of their holders beyond the actual scope of the patents (which is difficult to check) to deter potential entrants, notably small firms which cannot incur the cost of checking the rights and the possible ensuing litigation. The entry cost is also raised by the necessity for newcomers to negotiate with all patent holders, which could easily make the total cost of licences excessive (see below the 'double marginalization problem'). Patent thickets raise particular problems when there is no substitute technology available to the protected one and therefore cannot be circumvented.

The Use of Litigation or Threat of Litigation

Litigation is costly in terms of attorney fees, disturbing the attention of employees and managers (see e.g., Lanjouw and Shankerman 2004). This is all the more significant for SMEs, which are faced with stricter liquidity constraint and shortage of manager's time whereas the cost of litigation is a fixed one (it is the same for a small and a large company). Hence SMEs are particularly sensitive to threats of litigation and will make easier preys. They can be nearly ransomed by holders of patents of dubious legal validity, but would have to pay licensing fees as they cannot afford going to courts. The plaintiff may request injunction relief (stopping the production of the good allegedly infringing the patent), which can endanger the mere survival of the producer. Even large companies are subject to such threats, and often hesitate engaging into litigation for that purpose.

This explains the high activity of cross-licensing in certain industries, with deals often agreed as settlements to litigation. Firms that are not subject to such threats are those which are not engaged into production, such as the 'fabless companies' in the semi-conductors industry. Such firms have little to loose in litigation and much to gain. They have been named 'patent trolls', some of which have patent litigation as their only activity (acquiring patents from bankrupt firms, often at low price, and then threatening to bring the alleged infringers to courts). That activity has developed mainly in the US, as there is not yet a Europe-wide court system for patents, so that the cost of litigation is higher. Also, European courts are not as pro-patent as US ones (no damages comparable to the US are awarded) and it seems that more patents of dubious validity have been granted in the US than in Europe (Box 4.2. illustrates the strategic use of litigation threats with the RIM case, in the US and in Europe).

Box 4.2. The RIM case, part 1: infringer or a troll's victim?

Research In Motion Ltd (RIM) is a Canada-based manufacturer (Waterloo, Ontario) that launched its famous BlackBerry product in the US in 1999. In early 2006, the company had more than 4 million customers, the majority being based in the US.

US licensing company NTP, a small (about two employees) Virginia-based firm, sued RIM in 2001, accusing it of infringing at least five of its US patents on radio frequency wireless text communication. The patents were invented by Thomas Campana, Jr., an engineer who, in 1990, created a system to send emails between computers and wireless devices. Campana, who co-founded NTP, died in 2004. His wife owns a large stake in NTP.

The two companies settled at US$612.5 million for NTP in early March 2006. This was a full and final settlement of all claims until the five concerned patents fall into the public domain. It ended a long period of anxiety for more than 3 million Blackberry users in the US.

NTP was often called a patent 'troll' in the press, a firm that buys and owns patents in order to sue producers of successful products. Indeed, NTP was not involved in the business of wireless messages; made no investment in it; and apparently just waited until they could launch litigation against an infringer, in this case RIM, putting the latter and its shareholders at deep financial risk. NTP was winning its lawsuit and was therefore entitled to a permanent injunction—so forcing RIM to stop pursuing its business. The issue therefore concerned the total value (both the private and the society's benefit) of RIM's business.

The counter argument is to be found at the roots of patent systems, which clearly allow inventors to own, and sell, an idea. The validity of a patent portfolio cannot be related to whether its owner has or not material assets. It was stated in WSJ, 10 March: 'a patentee has no more duty to develop his invention than a farmer has to plough his field. He may license it to third parties or warehouse it to develop related products'. Companies that invest in patents but do not exploit them can be rather useful operators who act as intermediaries between those who create and those who commercialize inventions—i.e., market makers, not 'trolls'. Along this first line of argumentation, the lesson of the BlackBerry saga is that patents are there to be protected, and the settlement proves that the system works effectively, as it protects small inventors. Therefore NTP would not be such as a troll, but rather as a firm holding a patent portfolio that was infringed by another company.

The second line of arguments mitigates somehow this last conclusion. During the trial, RIM attorneys noted that the US Patent and Trademark Office (USPTO) was poised to finally reject all patents at the heart of the case. At the same time NTP withdrew all its patent applications at the EPO. Both the USPTO's re-evaluation of the several disputed patents, and the withdrawal of the EPO applications seem to suggest that the NTP patent portfolio should not have been granted in the first place, or should have been granted with a much smaller scope of protection. The issue is therefore related to the problem of policing bad patents. This might be done through an improvement of the rigour of patent offices, or through a weakening of the judicial presumption of validity. This second line of argument leads to the conclusion that NTP might be considered as a troll, as the firm successfully exploited a 'bad' patent . . .

The only way to draw a clear conclusion would be to assess whether (1) RIM purposefully used the patented invention to create its business and (2) the patented invention is really associated with a substantial inventive step, or whether it is 'obvious'.

In any case, it clearly seems that 'bad' patents are being issued, and that some companies use them to take advantage of this legal uncertainty, which basically means that it is possible to leverage a tax on innovators through the threat of patent litigation.

The RIM case is further developed in Chapter 6, Box 6.2, and Chapter 7, Boxes 7.2 and 7.3.

(See for instance *NYT* (2005) 20 December; *WSJ* (2006) 'Patently Absurd', 1 March; *WSJ* (2006) 'Troll Call', 6 March; and *WSJ* (2006) 'Letters to the Editor, Patent Law, Injunctions and the Public Interest', by Richard Epstein, Scott Kieff and Polk Wagner, 10 March).

Leveraging Patent Power by Tying a License on an Unpatented Product with a License on a Patented One

This type of licensing contract constrains the purchaser of a patented good (i.e., subject to limited competition) to buy an unpatented one (otherwise subject to competition). Hence the power given by the patent to its holder is extended to cover other goods than the one protected in the patent.

One example was the Trident case in the US, judged by the Court of Appeal of the Federal Circuit in January 2005. Trident manufactures ink jet devices, which are patented, and ink, which is not patented. The standard Trident's licence agreements grant the right to 'manufacture, use and sell . . . ink jet printing devices' to other printer manufacturers only 'when used in combination with ink and ink supply systems supplied by Trident' (in the court's words). According to the Appellate Court, Trident's licences are 'explicit tying agreement[s]' because the sale of a patented product is conditioned on the sale of an unpatented one. A similar claim was raised by the European Commission against Microsoft, which allegedly used its control over the Windows standard (not based on patents, but on copyright and network effects) to take over other markets, such as web browsers. It is also a growing trend in the pharmaceutical industry to sell a patented drug only as part of a combination pill which includes non-patented drugs mixed with the patented one, hence forcing patients to purchase a particular version of the unpatented drug (AARP et al. 2005).

Extending the Scope or Term of the Patent Beyond their Legal Values

The patent holders try to transform their exclusive right into a real monopoly on the market, e.g., by lengthening the duration of their control over the market beyond the patent life, or by buying out potential competitors. The former strategy consists in the filing, spread over time, of a series of follow-on patents protecting particular aspects or minor variations of the initial invention, which in fact keeps the initial invention protected for more than its own patent term. The latter strategy involves for the patentee to pay fees to other companies, in exchange for them to stay out of the market. These fees are in fact indirectly paid by customers, who are deprived of market competition and the corresponding reduction in prices. This is apparently practised in the drug industry, where it is a means for established companies to convince generic makers to postpone their entry into the market (see the *Federal Trade Commission of the US v. Schering-Plough*, 2005).

Grant Back Provisions

These are commitments by licensees to make available to licensors, for free or against payment, improvements they would make during the contract in relation with the licensed technique. The agreement can be mutual, with the licensor taking a similar commitment to the licensee. Grant back provisions are designed to facilitate licensing contracts under certain circumstances. If a company feels under the threat of being over taken by innovations based on its own technology, it will be more reluctant to license that technology, hence helping competitors. For the licensee, such a provision guarantees that it will not be licensed a second rate technique, as it can claim any improvement made by the licensor. The major concern of competition authorities with these provisions is that they reduce the incentive of both parties to do research for improving the licensed technique, as the successful party will have to share the reward with its partner.

Policy Implications

Some of these practices are clearly illegal, others are more 'abuses' of the system, permitting an undue extension of the exclusive right beyond the one granted by the patent office. These practices can be deterred, and often are, by a close monitoring by competition authorities. A question raised to patent offices is: to what extent could such practices be hampered upstream, by granting patents which would not facilitate, or would even hamper them? Could patent law and practice tackle some of these problems? It is clear that allowing titles to circulate which do not cover significant inventions facilitates such behaviour. It could be 'bad patents' (which are not really novel or with an excessively broad scope, or with a very small inventive step), or unexamined applications, with applicants taking benefit of slow procedures to make use of the uncertainty surrounding the yet unchecked rights.

Licensing is a crucial issue regarding the relationship between competition policy and IP policy. Many of the apparent frictions between the two types of policies are due to the perceived economic impact of licensing agreements, or symmetrically of the refusal to license. In which way can licensing agreements hamper competition? They can be used for sharing markets (by the inclusion of territorial exclusivity), or fixing prices (even indirectly). Some of these licensing practices are obviously illegal with regard to competition law. Particular types of provisions in licensing agreements, or even standard provisions in particular circumstances, can also reduce competition and hamper innovation in more subtle ways, by facilitating tacit collusion (see Regibeau and Rockett 2004 for a detailed presentation of such practices and their impact).

One argument in favour of licences is that they reduce the duplication of research. It saves on society's resources, but the drawback is that it reduces also competition between alternative technologies. Granting a licence is a way for inventors to avoid competition from alternative technologies and to expand their market share, which could result in inflated prices through higher licence fees. That raises the difficulty, in anti-trust cases, of identifying the relevant market that should be used as reference for inferring the market power of the licensor. Box 4.3. illustrates certain solutions found by European competition authorities in application of the Rome Treaty to mitigate the potentially negative economic effects of certain licensing practices.

Cross-licensing Agreements

These are contracts by which two or more parties grant to each other access to some of their patents, possibly with a complementary monetary transfer. They are thus a way for the companies involved to share some of their technology. Such contracts are often signed as part of settlements agreed by companies in order to avoid going to courts in a litigation procedure, which is extremely costly. Cross-licensing is a way of avoiding boundary wars between companies, in a context of unclear or extremely complex boundaries. In fact, many such agreements cover hundreds or thousands of patents. Patent holders, usually

---

Box 4.3. The EC Technology Transfer Block Exemption of 2004

The corner stone of the EC competition policy is Article 81 of the Treaty of Rome, which prohibits 'all agreements . . . which have as their object or effect the prevention, restriction or distortion of competition within the common market' (Article 81(1)). However, Article 81(3) offers the possibility of exemptions from Article 81(1) for agreements whose positive effects outweigh negative ones, in terms, e.g., of promotion of technical or economic progress. It is the purpose of the 'Technology Transfers Block Exemptions' (TTBE) Regulation (N. 772/2004) issued by the EC to explicit what such exemptions are in the field of licensing (the major type of technology transfer), hence to clarify what types of licensing agreements are permitted and what types are not.

The Regulation recognizes that licensing agreements 'will usually improve economic efficiency and be pro-competitive as they can reduce duplication of R&D, strengthen the incentive for the initial R&D, spur incremental innovation and generate market competition' (Recital 5).

The TTBE assumes that licensing agreements between competitors (including potential ones) pose a greater risk to competition than agreements between non-competitors. The block exemption applies when the combined market share of the parties does not exceed 20 percent of the relevant market if the parties are competitors, 30 percent otherwise.

The TTBE explicitly prohibits certain clauses in licensing contracts, the presence of which would make them void: such 'hardcore restrictions' include price fixing; reciprocal output restrictions; and limitations on the ability of the licensee to exploit its own technology or to do R&D.

The directive excludes from the exemption regime 'grant back obligations for severable improvements' (Recital 14), in other words, provisions which constrain the licensees to allow rights to licensors on improvements they would make on the licensed technique (grant backs) that could be exploited independently of that technique (severable).

The directive does not apply to cross-licensing agreements and patent pools.

large companies, know that much of their patent portfolio actually overlap, and they could be faced with endless legal disputes.

A second aim of cross-licensing agreements is to address situations where competitors could block each other, as each one holds patents which are necessary for a given product to be manufactured. The electronics industry is probably the most active in cross-licensing (Hall and Ziedonis 2001). The reason is the composite nature of many electronic products, which embody large numbers of complementary technologies, and the high propensity of companies in that industry to patent their inventions. Hence electronic products might require hundreds of patents, often held by different companies, for being manufactured (see Box 4.4.)

From an anti-trust point of view, cross-licensing can be viewed as a tool for collusion and as a barrier to entry. A tool for collusion because it can be used for supporting mutual threats between companies in order to monitor each other's behaviour (e.g., 'if you enter my market I will do the same to you, using the technology you licensed me'). Cross-licenses can be a barrier to entry as they create networks of intertwined rights and agreements (thickets) that make it more difficult for newcomers to enter the market, as they would have more difficulties finding a privileged partner among the incumbents who are all tied in with each other.

Box 4.4. Cross-licensing agreements in the electronic industry

In December 2004, Sony and Samsung announced that they had signed a cross-licensing agreement which would cover most of their respective patent portfolio—94 percent of Sony's 13,000 US patents and a similar percentage of Samsung's 11,000 US patents. These included patents covering basic technologies, such as DVD, DRAM, and flash memory chips, while 'differentiating patents' were excluded (those relating to Sony's PlayStation or liquid crystal displays (LCD)). 'With this agreement with Samsung we aim to keep clear of unnecessary conflicts and compete only in areas where we really need to compete', said Sony Executive Vice President, Yoshide Nakamura. Hence the goal was to halt growing legal tension between the two companies while preserving competition. The agreement included a payment of an undisclosed amount by Samsung to Sony, whose portfolio was deemed more valuable. The industry is accustomed to cross-licensing deals, but this one was unique in scale, with the high number of patents involved.

In April 2005, two other groups, LG Electronics of Korea and Matsushita of Japan, announced a similar deal. They settled a dispute over plasma display panel patents (each of them had filed a suit in Japan and Korea accusing the other of infringement). They were concerned that the dispute would spread to Europe and the US, hence weakening their entire respective market positions. In addition, LG Electronics is a major client of Matsushita for electronic parts, providing a further reason for settling the dispute. The cross-licensing agreement extended to DVD devices and personal computers.

(Reuters and other agencies' press releases: December 2004 and April 2005)

Patent Pools

These are essentially a large set of patents held by several firms: 'organisational structures where multiple firms collectively aggregate patent rights into a package for licensing, either among themselves or to any potential licensees irrespective of membership in the pool' (Clarkson 2004). In other words, two or more companies agree to put together their patents as a bundle that can be cross-licensed or licensed as a whole to third parties. It is similar to a single patent portfolio, and the fact that it has several owners is transparent to the licensee. Patent pools are useful as they reduce transaction costs. When a large set of patents held by different companies is necessary for implementing one technique or manufacturing one product, so that these patents are deemed essential and complementary, patent pools could make access to the technique possible at a lower cost.

The cost would be lower with patent pools because they mitigate the 'Cournot effect' (or double marginalization effect): if a set of licences from different companies are necessary for implementing one technique, each of these companies will behave like a monopolist, charging a higher price than the one that maximizes the collective gain. When a particular company increases its price, this has a direct positive effect on its profit and a minor, second order negative effect on the demand (as this company represents only a minor part of the market). But as all companies follow the same logic, the resulting set of prices will be so high as to reduce demand and reduce all participants' profits. This is a kind of negative externality, similar for instance to road congestion. Patent pools allow licensors to agree on a collectively rationale strategy, which maximizes their total profit while ensuring that the total cost for licensees will not be so high as demand would be depressed. Hence, patent pools are also in the interest of society as they reduce the cost of licences and enhance the diffusion of technology.

Anti-trust authorities are quite suspicious regarding patent pools. Patent pools were nearly banned in the US until 1995, and are now subject to regulatory clearance in the US and the EU. Patent pools in fact allow a set of players to take collectively full control of a technology area, which could easily be leveraged downstream, to the end customer, resulting in a monopoly. In order to prevent that, competition authorities have put a series of criteria to be fulfilled for patent pools to be authorized. The main one is that the patents that are included should cover complementary techniques only, excluding all patents covering substitutable techniques, which are essentially competing techniques. Including competing techniques in a single pool would restrict the choice of potential licensees, hence giving excessive market power to the patent holders (see Box 4.5.)

A second positive effect of patent pools is that they support the establishment of industry standards which in many cases might not be as sustainable or

Box 4.5. The 3G3P initiative

Concerns regarding patents covering the third generation (3G) mobile communication were raised in 1998. A very large number of companies, up to 100 according to certain sources, owned patents deemed essential to the realization of standardized 3G systems. The previous generation (with GSM) had been a bad experience for the industry (several litigations occurred) although there were only 20 holders of essential patents. Therefore, large companies and various public and private groups decided to set up a body which would facilitate the access to patented technology to all industry players at the lowest cost. This body is the Third Generation Patent Platform Partnership (3G3P), officially set up in 1999. The missions it was assigned by its founders are:

- to identify all essential patents in the field (hence reducing search cost and risk of litigation for its members); and
- to design standard licensing contracts that would reduce transaction costs and the total fees to be paid by licensees.

The group comprised both producers of equipment (e.g., Nokia, Alcatel, Motorola, and Sony) and telecom companies (e.g., BT and France Telecom). In view of the weight of such players, the 3G3P had to get clearance from competition policy authorities, in the EU, Japan, and the US. It received the green light in 2002, after having made substantial changes at the request of these authorities with the objective of limiting any possibility of collusion. The major change was the breaking up of the 3G3P proposal into five nearly independent entities, the PlatformCos (platform companies), each covering one type of technology (W-CDMA, CDMA-2000, etc.) The reason for the break up is related to the fact that the various technologies are competing with each other, and pooling them together would restrict competition.

Each PlatformCo is administered by a board comprising licensors only: licensees are excluded as their participation would facilitate collusion practices. Being a platform, not a pool, 3G3P provides common services and obligations to its members, but it is not as tight as a patent pool. The central platform is very light, providing only general rules that are then implemented in an independent way by each of the five PlatformCos.

What does 3G3P provide?

- An evaluation by independent experts of the necessary character of patents to 3G, resulting in a list of essential patents candidates for inclusion in the PlatformCo. The evaluation avoids members to make the search on their own, which is costly and subject to mistakes. Each member (including licensees) is supposed to submit to the Platform all its patents and applications that could be essential, including unpublished ones.

- Standard licensing terms are common to the five platforms (a single formula), but specific rates (fees) are established independently by each PlatformCo. These terms are used as default contracts.

- The licensing terms include the determination of royalties, intended to ensure that 'the total cost . . . is less than would be the case if essential patent holders negotiated individual licenses with everyone'.

- The licensing terms also include some kind of grant back provision, by which the licensees commit to put all their essential patents in the platform.

- The platform is not a pool: licences are still bilateral contracts (the licensee does not license at once all patents concerned by the platform), and the payment and collection of royalties remain the sole responsibility of the individual licensors.

> Hence the 3G3P appears as a kind of very sophisticated market for technology, with both constraints and freedom for participants. It tries to address issues such as the 'tragedy of the anti-commons' while keeping some margin for manoeuvre to companies, limiting the 'double marginalization effect'.
>
> (www.3gpatents.com)

as successful without the favourable licensing terms allowed by a pool. A standard that is backed by a patent pool can be less costly and gain market shares against non-standard (proprietary) technology. Many standards in the electronic industry are established on patent pools. For instance, this is the case of MPEG-2 for the compression of video data.

MPEG LA's MPEG-2 Patent Portfolio License was established to assure the interoperability and implementation of digital video by providing fair, reasonable, non-discriminatory access to worldwide patent rights that are essential for the MPEG-2 Video and Systems standard. Our MPEG-2 Patent Portfolio License makes worldwide MPEG-2 essential patent rights available to all users on the same terms at fixed rates under a single license . . . MPEG LA's goal is to provide worldwide access to as much of MPEG-2's essential intellectual property as possible; new Licensors and essential patents may be added at no additional royalty during the current term. Wide acceptance of the MPEG-2 Patent Portfolio License is responsible for the worldwide utility of MPEG-2 technology. The Program's Licensees make most MPEG-2 set-top box, professional (e.g., encoders, file servers and multiplexers) consumer electronics (including DVD player and television receiver/decoder), personal computer and packaged medium products in the current world market.

(http://www.mpegla.com/m2/)

The Setting of Standards

Standards are sets of technical specifications intending to provide a common design for products or processes issued by various providers (see Lemley 2002; Blind 2004; and Chiao et al. 2005 for a more in-depth analysis). Standards address issues such as interoperability, in other words, the ability of two pieces of technique (e.g., two machines or two software programs) to work together. Examples of standards include 3G (mobile phone) or Microsoft Windows. Standards are at the heart of many emerging industries, such as semiconductors, telecoms, and now biotechnology (e.g., with bio-chips). They are welfare improving as they allow to capture economies of scale in production, hence lowering cost and reducing the need for customers to use several different, possibly redundant, techniques in parallel. They are particularly important in markets that display network externalities, i.e., where a product is more valuable to a consumer if more consumers use the same product or a compatible product. Their social cost is, however, a reduction of technical diversity and the danger that a standard inferior to the best techniques would be chosen.

Standards can be set by market selection (competition between several alternative techniques supported by individual firms or groups of firms), by industry bodies (so called 'standard setting organizations', groups of companies specializing in certain technical areas and agreeing on technical design), or by governmental bodies (usually when safety is at stake). Standards can be open (accessible to all companies willing to implement them, such as the Internet) or closed (i.e., proprietary and so restricted to certain participants). The reality is generally somewhere between the two, with accessibility allowed under certain conditions (notably by payment of fees). Standards subject to ownership (and, in general, based on intellectual property) of individual companies or groups of companies are called 'proprietary'.

The impact of standards on competition is not clear cut. On the one hand, by levelling the playing field, standards enhance price competition downstream, notably when they are open. On the other hand standards may be used to exclude certain technologies from the market, or to reinforce the market power of their owner in the case of proprietary standards. Proprietary standards give a strong market power to their owner. In general, open standards are more favourable to competition and customers than proprietary standards, while the latter are favoured by companies in a position to control them.

Standard setting organizations apply various rules with respect to IPR. Some of them refuse to promulgate a standard with any IPR attached to it (that was the case until recently of the W3C, the body in charge of Internet standards) or require that their members give up such IPR. Such strictness is justified by the intention to have totally open standards. Other standard setting organizations (14 among the 21 surveyed by Lemley 2002) may impose limitation on the exercise of IPR, in the form notably of a pre-commitment by the holder to grant RAND licenses to any other member of the organisation or, in rare cases, even to non-members. RAND (reasonable and non-discriminatory) licences are licensing contracts that allow potential licensees to expect access to the technique at conditions that correspond to their economic needs. Some organizations may just require the disclosure by their members of IPR covering a standard prior to agreeing on such a standard. That obligation can be restricted to the already published IPR, or it can also extend to unpublished IPR.

The issues raised by IPR in their relation with standards are twofold. IPR can prevent standards that would be socially desirable, or standards can excessively weaken or reinforce the market power given by IPR. The first instance happens when a particular company holds IPR which is 'essential' to some standard (without which the standard could not be established), and will not accept to make the IPR available to other players at reasonable conditions. In view of the economic gains from standards, such a situation will happen only when the IPR holder is willing to impose its own, proprietary standard (e.g., for a PC operating system). It also happens that several players hold essential IPR: the danger then is the 'double marginalization problem' (two monopolists own complementary inputs, and each wants to price at the monopoly level, with a resulting price which is inefficiently high). By setting too strong conditions for the access to their IPR, they would prevent any agreement from being reached. However, it is clearly in the interest of all parties to reach an agreement, and in such cases the chances are that a patent pool would emerge to support the standard.

Standards might provide further market power to companies that abide to them, especially in the case of closed standards (i.e., when competitors are prevented from accessibility to the standard). This can happen if the holders of the IPR backing the standard refuse access to others at 'reasonable' conditions. Standards can also be used to leverage the market power of certain patents, as it has been alleged in recent cases (Intel in the 1990s and Rambus more recently). What is more, a company might simply not disclose its IPR at the time of standards negotiations, hence favouring the adoption of a particular standard (as all non-IPR holders would prefer an IPR-free standard), and once the standard has been adopted and broadly diffused, hence generating much market value, the holder would disclose and assert his or her rights. This particular type of 'hold up' (a manoeuvre by which an actor corners other actors after they have incurred the sunk costs they would have to write off if they refused the licensing offer) is known as a 'patent ambush' and is considered a form of competition abuse. The FTC of the US charged Rambus in 2002 with deceiving a standards organization of the memory chip sector, Jedec, with the aim of having one patented technology adopted as a standard. Then Rambus claimed royalties from chipmakers such as Hitachi or Toshiba. The case then went to the US courts as the EC was initiating an investigation of its own.

4.4. **The Value of Patents**

In view of the strategic importance of patents and/or licences to firms, the valuation of IP assets is a critical issue. Should a high tech start-up company be looking for capital (although financial valuation of patents plays no major role in most venture capitalist's investment decisions); or an established firm be willing to value another firm that it is about to take over; or, more simply, should any organization try and licence in or out its technology, the need for reliable valuation methods of intellectual assets and more specifically patents is critical. The growing importance of intangibles, and especially intellectual capital, on the balance sheet of many firms leads some of them to provide the value of their IP portfolio in their annual statements. Regulatory changes, such as the Basel II standards, require banks to valorize the risk associated with their loans, which applies notably to cases where the collateral is made of patents, whose value must therefore be assessed.

Beside these particular situations which require estimating the value of patents, patent holders and their competitors have to make frequent updates of the value of their own and of others' patents. Taking such decisions as whether or not to pay the renewal fees of an existing patent, to incur the costs of filing a new one or to designate states in which to extend its geographical scope do require prior informal valuation of the patent.

Valuing intellectual assets is all but a straightforward exercise for various reasons. First, the value of patents is highly volatile and idiosyncratic. It can change suddenly and unexpectedly with the obsolescence of the protected invention, due for instance to the appearance of a superior alternative, to shifts in consumers' demand, etc. Second, one should distinguish between the value of the patent itself and the value of the invention it protects. The former should be evaluated in terms of the value that it generates as compared to the value that would have been generated by the same invention without the patent; hence it is the value of the exclusive power on the invention, not the value of the invention itself. Third, the patent value depends as much on its contents as on its form. Poorly drafted patents may prove to be simply unenforceable, hence valueless, it being too easy to invent around them, despite the underlying invention being very promising and its patentability undisputable. Fourth, the value of a patent depends largely on the market strategy of its holder; ability to maintain its control over the market; and ability to improve the technique, etc. This factor is particularly important for small firms, whose skills are highly variable. It means that patent valuation is highly context dependent. Finally, the value of patents must usually be estimated before they get exploited. It is easier to value a patent that has already been exploited or licensed out to some third parties for a couple years: it is much more difficult to value it beforehand.

Subsequently, patent valuation methods, as any financial valuation exercise, generally imply some degree of forecasting hence uncertainty, should it be on future technological developments, future market conditions, firm capabilities or other relevant factors.

VALUATION METHODS

For lack of a straightforward and fully reliable valuation method, several methods are used by the industry, either to value one single patent or an IP portfolio as a whole.

Following Parr and Smith (1994) and Pitkethly (1997), valuation methods for a single patent can be organized into three main categories: cost-based, market-based, and income-based methods.

Cost-based methods, by far the most straightforward ones, rely on the assumption that the historical cost incurred in creating a patent is reflected into its value. By evaluating and discounting all the R&D and administrative or legal costs incurred while taking into account the effect of inflation, one may get an approximate value for the patent. Alternatively, another cost-based approach assesses the cost of replacing the patent or recreating an equivalent asset and using it as an estimate for the patent's value.

Although quite straightforward and objective, cost-based measures rely on a very hypothetic assumption, since the value of an asset does not always depend on the expenditures incurred to create it. What is more, identifying all costs purely and solely related to the investments into the patent can be difficult. Therefore, cost approaches should be used very carefully and only when expenditures or replacement costs can be reliably assessed. They require historical cost-based accounting systems. However, taxation authorities might request their use and they may be a relevant benchmark where a patent has recently been acquired (von Scheffer and Zieger 2005).

Market-based methods value assets by studying the prices of comparable assets which have been traded between parties at arm's length in an active market. This method is applied quite frequently in real estate valuation procedures. Such transactions may consist in sales of similar patents, but also on comparable royalty rates, should they be made of industry averages or based on R&D costs or sales.

The major shortcoming of this method is the uniqueness of patents and hence the difficulty to identify similar patents involved in recent transactions, and whose use represents the best use of the own patent being evaluated. In addition, the financial terms of private transactions involving patents are usually not disclosed, which reduces the pool of accessible information.

In addition to these 'classical' market approaches, an alternative method, described by Parr (1988), consists in deducting the value of the patented product of a firm from the market value of the firm subtracted of the book value of all known assets. This method can provide an approximate value for the patented product, although not for the patent per se. Furthermore, whether this residual value can be attributed solely to the patent remains largely arguable and such a method actually transfers the valuation power to the stock market, assuming that it has perfect information and methods to value the company's IP assets.

The third class of valuation methods consists in the translation of classical financial valuation schemes into the patent world, typically discounted cash flow (DCF) or net present value (NPV) methods. Basically, the idea is to forecast future income streams of the patent's commercial use and to discount these annual cash flows over the entire duration of the patent. Such methods should account for the elements of time and uncertainty in future cash flows. In order to further account for changing risks over time, some more sophisticated tools treat patents or patent applications as real options and use the Black and Scholes equation (Black and Scholes 1973) as the basis for valuation.

Like income-based methods, real option models require a valid initial estimate of the net present value of projected cash flows of the patent over its entire life from its inception to its lapse, including all the costs already incurred and yet to be incurred to file, maintain and enforce the patent up to possible opposition and litigation.

Estimating future cash flow is the major difficulty of income-based methods. If not readily available from real or forecasted cash flow directly associated with the patent, such estimates can only be based on industry averages of royalties paid by licensees for similar IP rights. But this would lead to the same shortcomings as the market-based approach.

Therefore, while the theoretical foundations and consistency of income-based patent valuation methods are superior to others for they focus on future earnings, they still require subjective cash flow allocations and the necessary information might not always be accessible from internal reporting systems.

From these methods for valuing patents as standalone assets, the overall value of a patent portfolio could be obtained by aggregation. But there is little certainty as to the form an optimal aggregation function should have, not to mention the enormous effort required to individually evaluate every single patent in a portfolio using the above methods, especially the income-based ones. This might be achievable for relatively small portfolios, but hardly for those that are large. Large patent holders would therefore probably prefer conducting internal surveys to get a broad overview of the distribution of the value within their portfolio, as illustrated for instance by Fröhling (2005), Head of Volvo's Patent Department.

PATENT-BASED FINANCIAL INSTRUMENTS

Nowadays patents are getting exploited in many different ways by their owners. From traditionally embedding their protected inventions into new products, processes or services, patent holders have now come more systematically to other ways of turning their IP into money, such as getting additional revenues from outward licensing, leveraging own IP assets to get access to external technologies through inward or cross-licensing, selling or purchasing patents to or from third parties, or using their patents as collaterals in capital sourcing.

An OECD survey (Sheehan et al. 2004) shows that 60 percent of the respondent firms reported increased inward and outward licensing during the previous decade (especially in the ICT and pharmaceutical sectors) and that getting revenues out of licensing has become the third most important reason for patenting. This phenomenon, illustrated by the creation of licensing spin-outs such as SBC Knowledge Ventures or BIPCO and of specialized licensing agents such as Thinkfire or IPVALUE, has been further examined in previous sections.

Increasingly also, patent holders are considering selling their patents to third parties. Should it be to get rid of a no longer core technology or of a broader scope of protection than the anticipated business (and then possibly negotiate to be granted back an exclusive licence on the field of use as part of the deal), to reduce the administrative burden of 'overhead' patents or simply to generate revenue and cash, patent holders may have various objectives in considering to give up some of their IP rights. Potential buyers may see in such purchases opportunities to fill some critical gaps in their current patent coverage, to obtain an initial patent base in a new business area, to get leverage for negotiating cross-licenses, or to simply purchase new assets to generate revenues from licensing or litigation. The latter objectives may be the strongest motivation of newly funded 'patent aggregators' such as Acacia Research or Intellectual Ventures, which are suspected of building patent portfolios mainly to extract settlements from producers or manufacturers.

New ways of exploiting IP rights allow their monetization through securitization, collateralization or sale-and-lease-back arrangements. Pioneered by David Bowie's record albums whose future royalties were monetized in 1997, raising US$55 million, IP securitization is increasingly seen as a promising way of generating cash-flow out of patents and other IP assets. Securitization is

a device of structured financing where an entity seeks to pool together its interest in identifiable cash flows over time, transfer the same to investors either with or without the support of further collaterals, and thereby achieve the purpose of financing. Though the end-result of securitisation is financing, but it is not 'financing' as such, since the entity securitising its assets is not borrowing money but selling a stream of cash flows that was otherwise to accrue to it

(Kothari 2003)

In its simplest form, securitisation consists in converting assets or cash-flows into marketable securities called asset-backed securities (ABS).

Thanks to their stability and risk-return characteristics, royalty streams from copyrights are especially attractive to investors. Patent securitization has lagged behind copyrights and trademarks, largely because of the volatility associated with the royalty streams, such as the risk that the patent will become technologically or economically obsolete (O'Haver 2003). Nevertheless, Yale University's success in 2001 in securitizing for Royalty Pharma a licence on its HIV-drug Zerit to Bristol Myers for a lump sum of US$115 million (Edwards 2002) shows the promise of such monetization deals, although the deal defaulted due to low sales. Royalty Pharma subsequently securitized a set of 13 patents to better diversify its holdings and mitigate associated risks (Hillery 2004).

The advantages of patent securitization to patent holders are numerous (Hillery 2004): the ability to obtain a greater amount of revenue than from a loan based on that future revenue; the fact that the capital is available upfront; a fixed interest for the duration of the deal; and the irrevocability of the bond sale which can become an insurance policy for the value of future royalties. What is more, in such transactions, patent holders retain the ownership of their IP.

Another financial instrument to leverage patents to raise capital consists in IP collateralization. The idea for the lender is to lend an amount based on the estimated value of the patent, depending on its quality and other risk factors. This allows patent holders to raise debt capital against the value of their IP rights without relying on the credit of their firm (Edwards 2002). An interesting case of such patent-backed loans is provided by Edwards in the form of GIK Worldwide's deal with Pitney Bowes Capital, which allowed GIK to raise US$17 million in debt.

Patent sale-lease-back arrangements provide technology-rich but cash-poor firms with yet another way of monetizing their intellectual property. The structure of such transactions is similar to real estate sale and lease-back ones: specialized institutions offer companies (sometimes their subsidiaries) to purchase, then licence back their patents, with a share of the proceeds returning back to the contributors over time. The initial patent holder enjoys a tax deduction for the amount of annual licence payments. The major drawback here is that he or she also loses ownership of the IP; however depending on the structure of the transaction, he or she may be granted an exclusive licence on the patented technique or an option for eventual repurchase of the asset at the termination of the licence period.

Finally, additional new financial instruments related to intellectual property rights have appeared in recent years in the form of patent litigation insurances. Large insurance brokers and underwriting companies are offering third party insurance products geared towards protection from infringement lawsuits. There are also insurance products to help finance infringement litigation for small companies that might lack the capital to pursue litigation as an enforcement measure (O'Haver 2003). But according to a study by CJA Consultants (2003) for the European Commission,

while schemes are marketed in the EU and the USA, including patent, trade mark and copyright cover, it appears that in no part of the world has Patent Litigation Insurance (PLI) been particularly successful, and more to the point, particularly in relation to SMEs, no insurance scheme has shown any capacity to provide adequate cover at premiums affordable by patentees in general. This is partly because of high levels of premium and low levels of indemnity.

Patent insurance, even more than standard insurance contracts, is confronted with moral hazard (the insurance takers might take more risk after they have the contract) and adverse selection (patent holders with the highest risk will take an insurance contract while those with lower risk won't).

All these new developments in the patent market have given rise to new specialized funds investing into this new asset class, essentially but not only in the US. Investment banks and merchant banks such as ICMB Ocean Tomo Bank have issued private equity funds focused on 'undervalued' patent portfolios. In Europe, IP Bewertungs AG has launched the first Patent Value Fund in

2004, which invests only in patents. The fund acquires patent portfolios from medium to large sized enterprises and—based on the portfolio's size—chooses either to individually customize an incubating fund or to design one directly as an institutional fund for the investor's individual requirements (Lipfert and von Scheffer 2006).

ECONOMIC APPROACHES TO MEASURING PATENT VALUE

In addition to the practical patent valuation methods in use in the industry, academics developed their own techniques for estimating the value of patents, and have investigated the determinants of that value.

One route for obtaining estimates of the value of patents has been to ask directly to the involved parties, the patent holder (Harhoff et al. 2002) or the inventor (Gambardella 2005), through opinion surveys. Hence, the Patval survey gathered responses from 9000 European inventors having had a patent granted by the EPO. They were asked to provide an estimation of the value of one of their patents, in terms of the minimum price at which they would have been willing to sell it on the day of the grant, with all the information they had on the day of the survey. These surveys have confirmed the fact that the value distribution of patents is highly skewed: many of them are worth very little, while a few are highly valuable. Such exercises rely on the assumption that patent holders or inventors have an efficient technique for calculating the value of their own patents.

Indicators of value can also be obtained from the patenting process itself (Lanjouw and Shankerman 1999). It is likely that a patent which is more often cited is more valuable (its importance is acknowledged in subsequent inventions), or a patent which is granted as opposed to refused (its quality is recognized by the patent office), or a patent which is renewed for a longer period of time (as it is more costly), or a patent which is filed in more countries (Lanjouw and Shankerman 1999), or a patent which is opposed (it hampers competitors' strategy). These indicators have been validated, by confronting them with the monetary value of the patent as estimated from surveys, or by correlating them with the stock market valuation of the company holding them (Hall et al. 2005).

Using these variables as indicators of value, it is then tempting to attempt to correlate them with other characteristics of the patents, which could then be interpreted as 'determinants' of the value. Hence it was shown that the number of co-applicants or co-inventors associated with the patent, the existence of international cooperation reflected in cross-border ownership of inventions and international co-inventions, all increase the value of the patent as measured by the probability of grant (Guellec and van Pottelsberghe 2000). Correlation could also be found with the institutional origin of the references cited in a patent (Sapsalis and van Pottelsberghe 2006a, b).

Economic approaches to patent valuation and business methods are increasingly combined into composite indicators, practically used in patent auctions, licensing deals, and some of the financial arrangements mentioned above (see for example Ocean Tomo's metrics, including the number of citations in other patents, the area of technology, and the number of claims, to try and provide patent holders and potential licensees a more accurate value estimate of a patent).

4.5. **Summary**

- Patent strategies are variegated and often sophisticated, with aims going far beyond the basic justification of patent systems.
- Strategic patenting (e.g., patent thickets, picket fences, inventing around, and defensive patenting) can result into a snowball effect of patent applications by the business sector.
- Licensing-out and licensing-in are more and more frequent practices, as witnessed by the growth of the market for technology, most probably of more than US$100 billion worldwide in 2006.
- Licensing deals are generally socially beneficial as they increase the diffusion of technology. But some anti-competitive practices, or abuses of the system, have led to a more intense intervention of competition authorities.
- The increased number of patents and their more sophisticated exploitation has strengthened the need for reliable valuation methods. The cost, market, and income approaches are the prevailing ones. All of them involve a certain degree of speculation.
- IP rights are increasingly used as financial instruments, including as collaterals to loans, as securitized assets, and as tradable assets.
- Economic approaches to valuation have shown that the value distribution of patents is highly skewed.

5 Patent Design

Dominique Guellec

The study of patent design investigates what inventions should be patentable or not, and which rights and duties should be attached with a patent. It addresses from an economic point of view what legal scholars call 'substantive patent law' as opposed to 'formal patent law' and includes aspects such as the patentable subject matter, the inventive step, and scope or duration of patents. These features could be considered as 'policy leverages' according to Burk and Lemley (2003), as they can be used by governments to influence innovation patterns in the economy. This chapter aims at understanding the major features of patent law, their economic rationale and consequences on innovation and diffusion patterns.

The chapter first addresses the question of 'patent quality' from the point of view of economics (first section). It then reviews the major instruments of patent policy that contribute to quality: subject matter (second section), inventive step (third section), scope (fourth section), and duration (fifth section).

5.1. **What is a 'Good Patent'?**

THE ECONOMIC NOTION OF PATENT QUALITY

The economic notion of 'quality' of a patent refers to the objectives of a patent system: to encourage innovation and the diffusion of technology. In this utilitarian approach a patent should be granted if and only if it is beneficial to society. The benefit of granting a patent is estimated from the comparison of a situation in which the patent is granted with two alternative situations:

1. There is no invention at all (i.e., there would have been no invention at all without the prospect of a patent).
2. The invention is made but it is not patented.

In addition, granted patents should be designed (in terms of inventive step, scope and duration) so as to maximize the benefit society can obtain form the invention. Therefore, analysing patent quality requires first to investigate the expected effects of patents on the inventor and on the economy. On the positive side, one will find:

- Higher profits for the inventor (as compared with its private return without patent), which would induce higher incentives to invent (the invention might not be done in the first place if it cannot be protected).
- Improved disclosure: an existing invention might be kept secret in the absence of a patent, which would impede the possibility of further inventions (of substitute products or of improvements over the original invention).
- Market for inventions: as tradable assets, patents facilitate the circulation of technology from producers (inventors) to most efficient users.

On the negative side, the possible social costs are:

- Deadweight loss due to reduced competition, resulting in higher prices and a lower level of consumption over the period of protection.
- Use of the new knowledge is blocked or made conditional on payment of a fee, which impedes its use in subsequent research, hence reducing the speed of technical change.

Patents should be granted only in cases where the positive effects are higher than the negative ones, and characteristics should be fixed so as to maximize the net positive benefit if a patent is granted.

The economic notion of patent quality differs from the legal and technical notions. The legal approach emphasizes 'legal certainty': a patent is of good quality when its probability of being upheld by the courts, in case of litigation, is high. The technical notion refers to the exhaustiveness of search and depth of examination: no relevant prior art should be missed, and the inventive step should be carefully assessed. Both notions refer to the proper implementation of the patent law and case law. The economic approach does not deny the importance of legal certainty and technical quality: they are even pre-requisite to economic quality, as legal uncertainty has negative effects on competition and investment, while patents of insufficient technical quality (e.g., not novel as some prior art would have been missed) undermine competition and increase business risk. But the economic approach goes one step further: it takes as the reference the missions of the patent system, and investigates not only the implementation of the law, but the law itself. A patent conforming to the statute, hence offering legal certainty and technical quality, could be seen as low quality from the economic point of view as it would not encourage innovation or would limit the diffusion of knowledge—due to a too low inventive step or too broad scope in view of market conditions.

By confronting the goals of the system with its possible effects, one can work out a more precise view of the properties of a quality oriented patent system. It should address the following questions:

1. Would the invention have happened without a patent? If the answer is 'no', then a patent should be granted (provided that the invention itself is beneficial to society). If the answer is 'yes', then one has to turn to question 2 below. It is of course extremely difficult to determine the conditions which will make a patent necessary or not for an invention to occur. Certain indications or indirect evidence can be investigated:

   • How costly was the invention as compared with business standards? In other words, was the invention obtained by chance, as a by-product of another activity, or did it require a considerable dedicated effort?
   • How risky was the research that led to the invention (risk is one component of the cost, ex ante)?
   • Are there alternative ways of appropriating the return from the invention (e.g., secrecy, advance on learning curve, copyright, government subsidies, etc.)?
   • Could a similar invention have come out of publicly sponsored research (e.g., universities)?

   If the invention was costly, risky, has no effective alternative way of protection, and is outside the interest of government funded research, then it probably needed the prospect of a patent, permitting a high mark up with significant sales.

2. How will the grant of a patent affect the diffusion of the invention? The issue concerns both the marketed goods based on the invention (direct use by customers), the knowledge, and the invention itself (use for further inventions).

   • Exclusivity: if the invention is an 'essential facility' (i.e., it has no substitute because it is difficult to 'invent around') then the patent gives its holder an extremely strong exclusion power that could impede diffusion, generating a high deadweight loss and delaying follow up research by third parties. Alternatively, if there are, at least potentially, equivalent inventions available, then a patent will give only limited exclusion power to its holder.

   • Commercialization: will a patent increase the likelihood of the invention being commercialized? This applies mainly to inventions by universities and by small companies which don't have the capabilities to manufacture themselves their inventions and need partners to do it: this contractual relationship, usually a licence, could be facilitated by a patent.

   • Disclosure: will the invention be disclosed if there is no patent? For process inventions for instance the answer is likely to be 'no', while for product inventions it can be 'yes'. Hence, the alternative to a patent is secrecy in the first case, while it is more like the public domain in the second case.

Overall, the effect of a patent on the diffusion of an invention is more likely to be positive (or less negative) when substitutable inventions exist, when the inventor does not have the capabilities to manufacture the invention, and when the invention concerns a production process, rather than a product.

Applying this series of tests to individual patent applications or to entire categories (e.g., technical fields) would make the patent system of high quality from an economic point of view. However, translating these criteria into an operational road map for law makers, judges and examiners is not straightforward. There are several reasons for that.

1. The two goals of the patent system are not independent of each other, they are even often conflicting. Increased disclosure for instance would make it easier for competitors to use the invention or invent around it, hence reducing the private return to the initial inventor. Conversely, strengthened rights of exclusion would increase the incentive to invent but would reduce diffusion. There is a trade-off here, and in certain cases only unambiguous conclusions will be reached. One instrument which makes the two goals concordant is policy aiming at developing markets for technology, which can altogether increase the diffusion of the invention and its profitability to the inventor.

2. Many of the conditions set above are beyond the reach of institutions governing the patent system as they are concerned with events occurring before the patenting procedure or after it. They would involve authorities in charge of research and innovation policies (regarding, for example, the availability of public funding for research), or authorities in charge of competition (regarding the degree of exclusion exercised by the patent holder). That points to the need for close coordination between the bodies in charge of the various policy areas.

3. The analysis above relies on characteristics of inventions which differ from one invention to the next. Criteria for patenting should be clear to all parties, rule-based, and predictable. It is the expectation of a patent (and a patent with specific characteristics) which influences the decision by a company to start research or not: hence the need for criteria which would be predictable at the outset. Much information requested to answer the questions above is available only late: the private as well as the social value of the invention will be revealed over the years. In addition, this information is not necessarily available at all parties concerned—companies won't reveal to outsiders the private value of their inventions (assuming that they know it), while the social value is a very elusive concept, often impossible to measure at all. An effective patent system must be based on information which is reliable, shared by all parties, and available at the time the decisions have to be made. It should also rely on rules, leaving as little discretion as possible to examiners in order to improve predictability and legal certainty.

The developments above points to interference of economics with patent laws and bodies: patent policy should be conducted in close coordination with other policies aimed at encouraging innovation; and economic concerns should be reflected in patent law. The former means that certain decisions which would make sense from the point of view of the patent system in isolation could be rejected in view of broader concerns, and vice-versa. For instance, certain patents which fulfil usual criteria could be rejected on competitive grounds—as they cover essential facilities or as they are part of a patent thicket built by the holder to block all competition. Patent institutions have no mandate for now to address economic issues. The use of economic criteria for granting patents is explicitly rejected by the EPO, as illustrated in the guidelines for examiners on the basis of decisions taken by boards of appeal: 'The EPO has not been vested with the task of taking into account the economic effects of the grant of patents in specific areas of technology and of restricting the field of patentable subject matter accordingly' (Guidelines to Examiners, Part C, p.49). That does not exclude, however, that technical and legal criteria could reflect certain economic concerns: it is actually often the case as we will see below.

Box 5.1. Sequential inventions

The patent system is essentially a market-based reward mechanism for inventors: the reward to the inventor is commensurate to the market success of the invention. What happens if an invention is not marketed? It happens notably when an invention or a set of inventions (upstream) is an input to another invention or set of inventions (downstream), which are then sold on markets. The upstream invention should obtain part of the market-generated reward, but how much? And how do patenting rules influence the sharing of revenue between upstream and downstream inventions? The issue is raised in particular areas such as genetic research (where gene sequences are an input to downstream research for cures), or in software (where many products are made of inventive bricks coming from various sources). This issue is investigated in economic models of 'sequential invention', or 'cumulative invention' following notably works by Suzanne Scotchmer (2004).

   The issue for society is to share the total revenue generated by the downstream invention so that both inventors (downstream and upstream) have interest to contribute, and to design the mechanism so that both will participate. The first point requires that each of them obtains revenue higher that his cost. The latter is more subtle: it requires that once one party has committed vis-à-vis the other to doing some operation, it could not cheat. For instance, the upstream inventor could make his or her invention available to the downstream inventor in exchange for a share in the revenue, but afterwards the downstream inventor would deny having used that upstream invention.

   The issues raised by sequential inventions to patent design can be addressed with various tools. The subject matter restricts the ability to patent pure discoveries, which are usually extremely upstream the inventive process. The inventive step limits the ability to patent a downstream invention which would not add enough novelty to the upstream one. The respective scope given to the upstream and downstream inventions determine the bargaining power of their holders, hence their respective share in the income generated by the two inventions. In addition, patent law offers special tools for dealing with such situations such as, a grace period, provisional applications, divisionals, 'dependent patents', etc.

5.2. **Subject Matter**

The major substantive criteria used by the EPO for granting or rejecting patent applications are defined in Article 52 of the European Patent Convention:

1. European patents shall be granted for any inventions which are susceptible of industrial application, which are new and which involve an inventive step.

2. The following in particular shall not be regarded as inventions within the meaning of paragraph 1:
    (a) discoveries, scientific theories and mathematical methods;
    (b) aesthetic creations;
    (c) schemes, rules and methods for performing mental acts, playing games or doing business, and programs for computers;
    (d) presentations of information.

3. The provisions of paragraph 2 shall exclude patentability of the subject-matter or activities referred to in that provision only to the extent to which a European patent application or European patent relates to such subject-matter or activities as such.

4. Methods for treatment of the human or animal body by surgery or therapy and diagnostic methods practised on the human or animal body shall not be regarded as inventions which are susceptible of industrial application within the meaning of paragraph 1. This provision shall not apply to products, in particular substances or compositions, for use in any of these methods.

(Article 52, European Patent Convention)

Paragraph 1 of the Article states general conditions on patentability, and paragraph 2 lists a series of exclusions which would not come from paragraph 1. It is through the substantiation of these conditions in particular circumstances that patent law operates.

The list of non-patentable new ideas in Europe includes artistic creations, scientific discoveries, metal acts, abstract ideas, business methods, software 'as such', animal and vegetal varieties, methods for medical treatment, and medical diagnosis. All jurisdictions in the world have such exclusions. The TRIPs (Article 27) states that, 'patents shall be available for any inventions, whether products or processes, in all fields of technology, provided that they are new, involve an inventive step and are capable of industrial application'. The text then lists a series of exceptions that is more restricted than the one in the EPC. In the US, patents could be granted to 'any new and useful process, machine, manufacture, or composition of matter' in the words of the US patent statute (35 U.S.C. § 101). That excludes notably laws of nature, natural phenomena, and abstract ideas. A Supreme Court statement of 1980, that 'anything under the sun made by man' is patentable, established utility instead of technicality, which prevailed before, as the basic test for patentability. That opened the way to a continuous extension of the patent subject matter in the US over the 1980s and the 1990s through a series of courts decisions, to include genetic material, software, and finally methods of doing business. A similar path was adopted in Europe, although more cautiously and not going as far as the US: genetic material and software-related inventions (or the so-called computer implemented inventions, CII) became progressively patentable over the 1990s and 2000s through a series of decisions by the EPO Board of Appeal.

Why is the field of patentability restricted? It is striking that patent laws give lists, but no theory: the reasons why certain subject matters are excluded is not explained. The theory followed by the EPO case law is based on the notion of 'technicality': all technical matters, and only them, should be patentable (provided that they fulfil other criteria). That theory raises two problems. First, its justification is not clear: why, say, should methods of playing cards not be patentable whereas chemical substances should be patentable? Second, it does not rely on an explicit definition of technicality, and in fact technicality is sometimes defined in a recursive way, as being the patentable subject matter after judges have decided on what is patentable and what is not. For instance, the extension of the patent subject matter to genetic material was done by reconsidering the boundaries between 'discovery' and 'invention' (technical), as material existing in nature (used to be seen as a discovery) was then considered as an invention if it was purified and extracted from its natural environment.

Keeping to the technicality approach has its limits as the frontier between technical and non-technical matter is not clear cut (there are management techniques, accounting techniques, artistic techniques, techniques for relaxation, etc.), and it is increasingly blurred with information technology allowing mental steps to be performed by a machine (e.g., a chip or computer) instead of the human brain, and with certain social sciences becoming increasingly grounded in mathematics and other hard sciences. One example is ergonomics: would software which improves the comfort of the user, allowing, for example, visual efforts to be reduced so helping the eyes, be considered technical? It has clearly an effect outside the computer, this effect could be measured (visual effort and attention can be objectively tested). The technicality theory cannot address properly such changes in the boundaries of knowledge, which are more and more frequent. In view of the blurring of the notion of technicality, certain jurisdictions like the US are progressively opening the door to universal patentability, patentability of all types of man-made creations. This makes sense in the absence of any meaningful theory which would prescribe the proper limits to the system. Such a theory can be found, under the condition of turning to the policy missions of the patent system. The economic and social effects of patents should be the guiding principle: This might not always give unambiguous responses, but it would at least suggest empirical testing, making it possible to have evidence based policy choices.

Why should patentable subject matter exclude certain types of man-made creations? In the words of the UK High Court: 'Excluded items /of article 52 EPC/ . . . were excluded for policy reasons' (UK High Court 2005). Accordingly,

---

Box 5.2. The protection of art

Aesthetic creations are protected by copyright, a legal system that covers the *expression* of ideas, not ideas themselves (contrary to patents), then allowing free use and circulation of ideas by all, while preventing mere imitation (aesthetic creations can also be protected by laws covering industrial design or models). Leaving ideas free to circulate is a direct response to concerns with freedom of expression, but it has also some economic meaning. Most inventions in the arts are concerned with expression more than substance (the 100+ different ways of painting a landscape). Whereas secrecy is an alternative for technical inventions, it is of course not viable for aesthetic ones, whose contribution and value is in the visible features of the creation: hence no incentive to disclose (one motivation for patents) is needed there. Granting protection to the idea of painting a landscape would have impeded progress in the arts as it would have hampered the diffusion, hence the improvement, of the idea. For the arts there is also a problem of description. One patent on the *Mona Lisa* for instance would cover her special smile: but how could it be described in an unambiguous way, so that any infringement could be identified? Although not always straightforward, the description of technical inventions is easier. Finally, as the protection given by copyright is narrower than the one provided by patents, the duration of the exclusive right has been extended so that the revenue appropriated at the end could still be substantial (it was not always the case: until the beginning of the nineteenth century copyright was shorter than patents in many countries).

---

economics recommends two tests for deciding on patentability of a field: (1) will patents encourage inventions? And (2) will patents encourage or impede the diffusion of technology? Fields should be excluded from the subject matter when it is likely that patents are not needed for encouraging innovation or when they might have too detrimental effects on the diffusion of technology.

SUBJECT MATTER AND INDUSTRIAL APPLICABILITY

The only general, transversal criterion for restricting the patent subject matter which is given in the EPC is 'industrial application'. Article 57 of the EPC, complementing Article 52 in that regard, states that 'an invention shall be considered as susceptible of industrial application if it can be made or used in any kind of industry, including agriculture'.

Rule 27(1)(f) of the EPC prescribes that the description should 'indicate explicitly, when it is not obvious from the description or nature of the invention, the way in which the invention is capable of exploitation in industry'. The case law indicates that the notion of 'industry' has to be interpreted broadly to include all manufacturing, extracting and processing activities of enterprises that are carried out continuously, independently, and for financial (commercial) gains (see, for example, decision T 144/83 OJ EPO 1986).

'Industrial application' has often been interpreted in Europe as meaning an application in the *manufacturing* industry (and agriculture), adding weight to the notion of technicality seen as synonymous to materiality. It is a classical and restricted view both of industry (it excludes service sectors, now 70% of GDP of developed economies) and of science (as quantum theory has 'dematerialized' matter and put information on the same level as energy for being a basic component of the universe). This interpretation is opposite to the US approach, which relies on the notion of 'utility', which in particular does not exclude inventions of non-material nature (e.g., in the field of computing).

The criterion of industrial application is also a way to keep out of the patent subject matter basic, upstream inventions which could have various down-stream applications not yet well worked out. Keeping such inventions in the public domain (assuming that they are not kept secret) allows further inventors to use them for finding applications, and it gives them incentives to do so. It is only once the industrial applications are marketed that society benefits from the invention. The 'reduction to practice' obligation is a way to make it more attractive to do downstream research, although it could be at the expense of upstream research (see Box 5.1 on sequential inventions). In addition, patents on upstream inventions would probably be extremely broad as they would implicitly cover all downstream applications not yet worked out: think of patents possibly granted to Maxwell for work on electromagnetism, or Leonardo da Vinci on the helicopter, contact lenses, etc. A similar issue arises in emerging technology fields, which rely heavily on basic science and which still advance with radical inventions. Biotechnology and software are modern examples of that phenomenon.

BIOTECHNOLOGY

The patentability of genetic material was established progressively, through decisions by courts, including the Supreme Court in the US (in 1980) and the board of appeal of the EPO (in the 1990s). Technical arguments were used for establishing the patentability of genetic material. It was deemed that, once it is isolated from the body, genetic material could be considered as an invention, not a discovery. Opponents claim that living material is a discovery, not to an invention, as it exists in nature before being identified by researchers. For plant and animal varieties, which are explicitly excluded from the patent subject matter (EPC Article 53(b)), the Enlarged Board of Appeal stated that patents could be granted for genetic modifications that apply to more than one variety (decision G1/98), as they go beyond the traditional work of breeders (who experiment at the level of individual varieties only). The whole issue was clari-fied by the biotech patents directive (98/44) voted by the European Parliament in 1998. Article 5 of the directive, allows the patenting of 'an element isolated from the human body or otherwise produced by means of a technical process . . . . The sequence or partial sequence of a gene may constitute a patentable invention, even if the structure of that element is identical to that of a natural element.'

To investigate the issue from an economic perspective, we apply the two basic questions: Are patents useful for making inventions to happen in the first place? And what impact do they have on diffusion and on further research? Regarding the first question, genetic research used to be the realm of universities, sponsored by government, and of pharmaceutical companies which would be rewarded by new drugs they would make thanks to these basic inventions. Patenting of genetic material allowed the emergence and expansion of the biotech industry. Many companies in that industry have no asset but their patents. Many do not have the capabilities to go downstream, doing clinical tests or manufacturing drugs: they earn revenue by licensing or selling their patents to pharmaceutical companies, or by being taken over by these companies, which thereby acquire their patent portfolio. Those which go to the industrial stage (such as Myriad Genetics, offering testing for breast cancer) rely heavily on their patents, without which they would soon be imitated. There are little questions that without patents much genetic research would have happened any way, done by universities (government funding) or by vertically integrated pharmaceutical companies. The issue is whether it would have been conducted at the same scale and as efficiently as it has been. No definitive answer is possible. One could note that the biotechnology industry is extremely competitive, more so than universities or big pharmaceutical companies, which might have accelerated research in the field. That was clear in the case of the decoding of the human genome, where competition between the public consortium (the Human Genome Project, HGP) and a private company (Celera) led to the invention of new techniques which allowed faster decoding of genes. Patents have also provided university teams further incentives to perform their research and to frame it in a way that would facilitate downstream applications. As for pharmaceutical companies, biotech is a disruptive technology and one can guess that they might not have had as strong motivations as biotech startups to conduct research so rapidly in this area, where results would destroy part of their business.

Regarding diffusion and downstream research, the impact of patenting of genetic material is ambiguous as well. Certainly patents erect barriers to access. Normally, a licence granted by the patent holder is necessary for carrying out research making use of the protected genetic material. Hence the fear that transaction costs and double marginalisation (see Chapter 4) would block downstream research in the field. There are cases were it took time and effort for researchers to assemble the set of access rights they needed for doing their research (e.g., the vaccine for malaria—see Garrisson 2004). To a certain extent these barriers could be circumvented through licensing agreements, research exemption or mere ignorance of the patent by the user. These mechanisms seem to have worked reasonably well in the US (Walsh et al. 2003). Even if difficulties to use patented knowledge for research purposes do not seem to endanger research at a large scale, there is some evidence that they do exist, for instance in the US for human stem cells (which are not patentable in Europe) and are growing, even for research in the industry (Hansen et al. 2005). On the other hand, patents involve publication of the protected material (i.e., the entire genetic sequence should be disclosed). One alternative to patents is government sponsored research, which would trigger disclosure as well. But the other alternative is in-house research by vertically integrated companies, which would not disclose their upstream inventions for fear that their competitors could exploit them better of more rapidly than themselves. There is evidence of such protective strategies occurring for the decoding of proteins (which are stored in proprietary databases). In fact, for fear that biotechnology

---

Box 5.3. Scientific result v. industrial applicability (extract from Decision T 0870/04 of the Board of Appeal of the EPO)

---

1. Merely because a substance (here: a polypeptide) could be produced in some ways does not necessarily mean that the requirements of Article 57 EPC are fulfilled, unless there is also some profitable use for which the substance can be employed.

2. For the purposes of Article 57 EPC, the whole burden cannot be left to the reader to guess or find a way to exploit an invention in industry by carrying out work in search for some practical application geared to financial gain without any confidence that any practical application exists. A vague and speculative indication of possible objectives that might or might not be achievable by carrying out further research with the tool as described is not sufficient for fulfilment of the requirement of industrial applicability. The purpose of granting a patent is not to reserve an unexplored field of research for an applicant.

3. In cases where a substance, naturally occurring in the human body, is identified, and possibly also structurally characterized and made available through some method, but either its function is not known or it is complex and incompletely understood, and no disease or condition has yet been identified as being attributable to an excess or deficiency of the substance, and no other practical use is suggested for the substance, then industrial applicability cannot be acknowledged. Even though research results may be a scientific achievement of considerable merit, they are not necessarily an invention which can be applied industrially.

(p.1)

In the board's judgment, although the present application describes a product (a polypeptide), means and methods for making it, and its prospective use thereof for basic science activities, it identifies no practical way of exploiting it in at least one field of industrial activity. In this respect, it is considered that a vague and speculative indication of possible objectives that might or might not be achievable by carrying out further research with the tool as described is not sufficient for fulfilment of the requirement of industrial applicability. The purpose of granting a patent is not to reserve an unexplored field of research for an applicant.

(p.20)

companies would patent SNPs (single nucleotide polymorphisms), large pharmaceutical companies founded in 2000 the SNPs consortium, which identifies these entities and put them in the public domain, in databases accessible on the internet. It is likely that if there had not been competition by heavily patenting biotech companies, pharmaceutical companies would have worked on SNPs in the traditional way, by just keeping their findings secret and proprietary. Patents then had the indirect effects of accelerating research and of enhancing the public domain.

Biotechnology illustrates a situation that might become more frequent in the future: pure science having foreseeable business applications. The 'natural' boundary between pure science (associated with discoveries, scientific theories, and mathematical methods) and technology has been in close correspondence in the past with the institutional boundary between university and business research: universities did science and published it, while business did applied technology and patented it. Hence the technical and legal justifications for keeping science out of the patent subject matter essentially corresponded to the economic logic—patents going hand in hand with technology and with business, and publication going with science and universities. In the past two decades both frontiers (between science and technology on the one hand, and between universities and business on the other) have been blurred and they do not match any more with each other. In many cases pure science, such as decoding the human genome or finding new primary numbers, has foreseeable technical and market applications, which means that they are increasingly patentable. Universities tend to get closer to commercial activities (e.g., doing contractual research with companies, creating spin-offs, etc.) With further advances of the knowledge economy, involving denser and shorter ties between advances of knowledge and market exploitation of inventions, that situation will become more and more common (in biotechnology, in information technology, in nanotechnology or cognitive sciences) and it will require the patent system to adapt.

SOFTWARE

A second issue for courts and public debates since the early 1980s has been 'software patents' or 'patents on computer implemented inventions' (CII). A CII is defined as 'an invention whose implementation involves the use of a computer, computer network or other programmable apparatus, the invention having one or more features realised wholly or partly by means of a computer program' (Art 2(a) of the Proposal for a European Directive on the Patentability of CII). Patenting of a computer program is explicitly excluded by the EPC. The exclusion however applies only to computer program 'as such' (Article 52(3)). The words 'as such' in that context have been interpreted by the

EPO Boards of Appeal in a series of decisions in the direction that computer code is not patentable (it is covered by copyright already), but the underlying inventions (CII) themselves are patentable when they have a technical character and fulfil other statutory criteria. The 'technical character' is defined as 'teaching for technical action, solution to a technical problem.' The term 'technical' (which is not defined in the EPC) is interpreted as 'having an effect beyond the normal physical interaction between a program and a computer'. For instance, processing physical data is technical but processing financial data is not technical. Physical data may be, for example, data representing an image (decision T 208/84 of the Board of Appeal of the EPO) or data representing parameters and control values of an industrial process (T 26/86). However, monetary values (T 953/94), business data (T 790/92), and text (T 38/86) are not physical data. The tendency here, like for the interpretation of 'industrial application' is to keep close to materiality, keeping patents for inventions with a material, tangible effect, and excluding applications or effects of an intangible nature, often associated with the provision of services as opposed to goods. That approach notably rules out patents for word processing, spreadsheets, and general business software (see Laub 2006). The justification is that all inventions, and only those ones, which are patentable when implemented by a physical device should be equally patentable when implemented by a software means. The way an invention is implemented does not matter, it is a commercial choice left to the patent holder; what matters is just the invention itself. Hence, a brake system related invention is technical and should be patentable whatever its means of implementation, whereas a financial method is not patentable, whatever its means of implementation.

Much debate has revolved around the delineation of technical arts, as opposed to other realms of knowledge such as science, mental acts, etc. Philosophical considerations have been invoked (e.g., in debates at the European Parliament), stating for instance that only inventions relating to 'forces of nature' (restricted to matter and energy, in accordance with German law) should be patentable. According to that logic, all types of information, including algorithms (and computer programs are mathematical algorithms running on a computer) should be excluded from patentability on the same ground as scientific theories, because they are abstract, kinds of 'automated mental acts', and they are 'discoveries' rather than inventions. An opposite view emphasizes that modern science (e.g. quantum physics) and information theory considers information as a 'force of nature', which is as necessary as the notions of matter or energy for describing the universe.

Economists would consider different questions, focusing on the impact of patents on the incentives to invent and on diffusion. Much debate has revolved around the alleged specificity of software as compared with other technical fields—is software so different that the negative effects of patents would be magnified while the positive ones would be reduced?

The technical specificities of software include notably 'incrementality', 'modularity', 'complementarity', and 'granularity' (e.g., Bessen and Maskin 2000). Software progresses by small steps contributed by many different researchers. Each contributor benefits from contributions by others, as they improve a product used by all. For the process to go smoothly, one needs easy access to computer code (so that it could be cross-checked and modified), and technical compatibility so that the various contributions can fit together. In that context, patents would erect barriers to access that would hamper the entire innovation process. Hence the conclusion: patents are detrimental to software innovation.

In that matter, one needs to distinguish between the production of new software and invention. Software production probably has the specificities mentioned above. One software piece could be produced by assembling existing sub-pieces or variants of them. This is mainly a matter for copyright, which applies to the distribution of software. But invention is a different matter: there is lack of clear evidence that the inventive process in that field differs from other fields. For instance, the incremental nature of knowledge progression is universal ('we are dwarfs standing upon the shoulders of giants' as Bernard de Chartres famously wrote in the twelfth century), most inventions in all fields are combinations of existing knowledge and it has not been demonstrated that software would be different in that regard.

A related argument is that, as software inventions are abstract, patents granted on them necessarily need to be broad, thus hampering downstream applications. In fact technology, in all emerging fields, is increasingly based on science, be it genetics or mathematics, and it is ever more 'abstract' (i.e. embodied in formula: letters of the DNA code, computer code, etc.) Banning patents on all matter having some degree of abstraction would exclude the patent system from the most dynamic technologies, restricting its relevance to the most traditional technical fields—mechanics and chemistry. A report from OECD shows that in the late 1990s more than 30 percent of R&D in all industries was related to software (Khan 2001), a figure which is certainly still increasing.

The production of certain software is organized a different way from production and innovation in many other fields: a decentralized, modular way, which allows the participation of many independent entities, possibly individuals. This is made possible by two factors: the entry cost to writing software is much lower than to most other economic activities, as it requires no more than skills and a computer; and independent programmers can put together their respective contributions by using the Internet. As a result, a community of software programmers emerged, made of users turned producers and of volunteers, organized in extremely well structured projects. That community works following an open source way of coordination—each programmer makes available to others the source code he or she writes, so that the various pieces could be arranged together and could be scrutinized by others. Open source software is diffused under various types of licences, generally labelled 'copyleft' (as opposed to copyright), which essentially give the right to the user to modify the code, add code, etc. provided that he or she makes this code available under the same conditions to other users. The Linux operating system and Apache web-server system are major examples of open source production. Motivations for individuals to participate in that community are related to training, to building professional reputation, or to improving the tools they use. These are not standard market incentives (e.g., wages, profits, etc.) However, in the strategic context of the software industry, certain large companies (e.g., IBM, Novell, and Sony) have joined forces with open source, giving it a foot in mainstream market competition and providing a very significant infusion of capital. The issue is the economic model of inventing and producing software. A significant portion of the software industry obtains resources from the sale of downstream, customisation services, from venture capitalists, and using voluntary, free labour of a significant number of highly skilled volunteer programmers. This part of the industry usually does not make use of patents (or for purely defensive purposes). However this model is not IPR free: it depends on copyright, which guarantees that no programmer modifying an open source program could appropriate the entire program—copyleft relies on copyright.

There are two other alternative economic models for the software industry in addition to the open source. The first one, 'proprietary software', exemplified by Microsoft, has prevailed in the 1980s and 1990s. The appropriation of a rent occurs through a strong control of the market thanks to network effects and other economies of scale-like mechanisms, helped by the secrecy permitted by copyright law (which forbids decompilation of software). The second model replicates the situation of other industries, relying (at least partly) on patents for protecting significant inventions while using copyright for preventing mere copying. How do these three economic models compare to each other in terms of efficiency on the various segments of the software market? Can these models co-exist, and, if yes, under which conditions—or are they exclusive of each other? We must unfortunately confess ignorance as little empirical research has been devoted to these questions, although answering them would go a long way in assessing the economic desirability of patenting software.

Another issue, not specific to software but particularly important in that area, is interoperability and standards. The various software programs used on a given machine, a computer or a network, need to be compatible with each other. For that purpose, they should comply with certain standards. A software producer with a strong market position, due to a high market share or an essential software piece (e.g., an operating system) can impose its standards, limit the access of other producers to these standards and use that for competing (unfairly) with them so as to expand its dominance. This has been the strategy allegedly followed by Microsoft, starting from the control of the operating system and extending it progressively to office software, multimedia, etc. The economic question is: how is interoperability best handled, with patents or without patents? The usual 'patent trade-off' applies. Open standards allow the supply of a broad diversity of goods offered by competing firms: this process can be hampered by patents. At the same time, good standards often imply substantial investment in R&D, which companies will not incur without the perspective of appropriating some of the benefits the standard should generate.

Open source without patents could be the ideal solution to both objectives (Bessen and Maskin 2000), as the publication of the source code makes the standard open, while the absence of protection encourages all players to adopt it, which benefits all of them. The latter effect is due to an expanded consumers basis, hence revenue, and to an acceleration in technical change (coming from the increased number of firms on the market). In that context, patents modify revenue sharing in favour of patent holders and reduce the overall size of the market as there is less competition and diversity. However, the absence of patents could have the opposite effect to favour a model based on proprietary and secret standards, protected by copyright, as it has prevailed in the market for operating systems. Patents at least involve some disclosure and facilitate licensing, cross-licensing, and various types of platforms so that standard can be shared (see Chapter 4).

It is sometimes argued also that software patents would favour big companies against SMEs, hence reducing competition and market entry, at the detriment of innovation and of customers. It should be noticed first that Microsoft, the company with the strongest grasp on its market in the industry, has its market power based on other grounds than patents—network effects, copyright, and secrecy. In addition, Microsoft was condemned to pay high royalties and damages to small business as the courts found it guilty of having infringed their patents (two famous cases are Stacker in 1993 and Eolas in 2005, which both claimed the invention of important components then integrated into Windows). The strong support given by Microsoft and other large companies to patent reform in the US in 2005 was aimed at cleaning the field of 'patent trolls', small companies (without factory; often just attorneys) which use their patent portfolio essentially for extracting licensing fees from actual producers under the threat of litigation. There is no evidence that software patents in the US have worked against SMEs (Mann 2004). On the contrary, patents provide SMEs with the only protection they could have, as they lack market control or brand name that largest companies enjoy. They could then be rewarded for their inventions, obtain licensing contracts, and raise capital.

Overall, before in-depth studies of the impact of 'software patents' are done, it does not seem that they have had visible negative effects on the industry as a whole, nor has a positive effect been demonstrated. There was much innovation in the software industry as soon as in the 1960s and until the 1990s without patents, while the lead of the US in terms of patenting software has not weakened the position of its industry as the worldwide leader. More immediate issues are about the quality—novelty, inventive step, scope, clarity of disclosure—of software patents, which has been characterized as extremely poor at least in the US.

BUSINESS METHODS

Business methods were long considered as not patentable all around the world. The 'State Street' decision by the US Court of Appeal of the Federal Circuit (CAFC) chaired by Judge Rich in 1998 changed all that. The CAFC upheld a patent on portfolio management granted to a bank (U.S. Patent No. 5,193,056). From that point, there was an explosion in patent applications for methods of doing business in the US, which then flooded other jurisdictions including Europe. The State Street case was still about a business method involving a technical feature—the use of a computer network. In October 2005, the Board of Patent Appeal and Interference of the USPTO, in the Lungren decision, held that no connection with the 'technological arts' is needed for an invention to be patentable: pure business methods would then be eligible to patenting.

The EPO Board of Appeal has ruled on several cases and has evolved a more restrictive position than the USPTO. In the 2004 'Hitachi decision' (T 0258/03) the Board held that

What matters having regard to the concept of 'invention' within the meaning of Article 52(1) EPC is the presence of technical character which may be implied by the physical features of an entity or the nature of an activity, or may be conferred to a non technical activity by the use of technical means. In particular, the Board holds that the latter cannot be considered to be a non-invention 'as such' within the meaning of Article 52(2) and (3) EPC. Hence, in the Board's view, activities falling within the notion of a non-invention 'as such' would typically represent purely abstract concepts devoid of any technical implications . . . For these reasons the Board holds that, contrary to the examining division's assessment, the apparatus of claim 3 is an invention within the meaning of Article 52(1) EPC since it comprises clearly technical features such as a 'server computer', 'client computers' and a 'network'

(EPO 2004)

This decision says that any claim with a technical reference (e.g., the use of a computer) should be considered as eligible to a patent provided that other criteria are fulfilled. The Hitachi decision however also made the case that when it comes to the inventive step, only technical features should be considered, not for instance the business method aspects. Hence, if the only technical feature in the invention is the use of a computer, this is not new and no patent will be granted.

Why would society exclude methods of doing business from the subject matter? From an economic perspective, this question is about the effect of patents on the invention and diffusion of new business methods. Are patents a useful addition to incentives to innovate in that area? Although no counterfactual is possible, it is noticeable still that there have been inventions in that field for centuries without patents. New methods for accounting or new business ideas have flourished without the additional fuel of patents. Many business methods applications in the US consist in putting on an electronic system, generally the Internet, methods which existed already in the 'real world', such as rules for conducting auctions. So, the *combination* of the traditional method with Internet is the novelty. When such novelties are inventive software, they qualify as patentable subject matter under the software category. Otherwise they are pure business methods and should be examined as such.

There are several reasons why patents would not create significantly stronger incentives to invent in that area. First, the cost of inventing new business methods is generally low. Most business methods are merely ideas, with some elaboration, but nothing comparable to regular research and development expenditure. Most of the cost is incurred at the implementation stage (e.g., building brand around the new product) and does not qualify to patent protection anyway. Second, business methods could be protected by many other means, such as first mover advantage, secrecy, trademark, etc. Patents do not add much to the reward accruing to the inventor.

A few empirical studies have shed doubts on the interest for society of patenting business methods (Encaoua et al. forthcoming). A study by Tufano (1989) on the incentives to innovate for investment banks between 1974 and 1986 showed that although lead-time was relatively short for innovators, they usually enjoyed durably higher market shares than followers. It is through this market share, not through underwriting spreads (mark up), that financial innovators are rewarded. Consistent with these empirical findings and additional information from interviews with credit derivatives dealers and developers in investment banks, a theoretical model developed by Herrera and Schroth (2000) concluded that the incentive of investment banks to launch new financial products comes from the informational advantage provided by being the first in the market. Lead-time is important because it provides an informational advantage in a context where learning by doing is crucial, rather than due to the short-term monopoly profits it may confer to the innovator before its competitors enter the market.

In fact, Herrera and Schroth (2000) argue that the threat of imitation prevents the innovators from charging monopoly profits in the learning stage, when they are the only supplier. The innovators are able to recover investment costs because their informational advantage and expertise enable them to have lower current costs and enjoy larger market shares than their competitors at the competitive stage. Information is essential in a market where profit depends on risk management. Although the design of a financial product is fairly easy to imitate, its optimal exploitation can only be imperfectly imitated because it requires some expertise that is carefully protected by the innovator with secrecy and trademark. One of the persons interviewed by Herrera and Schroth (2000), a J.P. Morgan credit derivatives trader, explained that the economic value of a financial product lies in information about its performance, rather than in its design:

Everybody can see the laid-out contract but what I am very careful not to disclose are the positions in my book. With this information you could track down the logic and see where I make money. Without it you could not price correctly the product, break down the risks involved, and understand what the components are. New ideas are not easily imitated: the developing process is a set of complex skills that are not easy to acquire.

These cases also show that patents do not add much to disclosure of inventions, hence to diffusion. First, for many business methods publicity happens in any case as soon as the new product is marketed. Second, business methods which can still be kept secret after a few months would probably not be patented, hence not disclosed anyway. It seems also that for business methods more than for technical inventions much of the cost of innovating comes with implementation rather than with the invention itself. Implementation is not based on a small set of codified and transferable actions and pieces of knowledge, but on much broader and tacit factors, often labelled under the heading of 'talent', 'organizational capital', etc. This is exemplified by Wal-Mart, renowned as an extremely efficient company for managing flows and stocks of goods, which actually holds a couple of patents, but whose methods could be only partially implemented by its competitors (Toyota holds a similar position in the automobile industry as Ford did in the early twentieth century, and Dell does now in the computer market).

Hence, there is probably not much incentive or diffusion effect to be expected from patents in the area of business methods. Not only patents might not encourage inventions, but they could create trouble. Inventing new ways of doing business is the essence of market competition: new ways of attracting customers, more efficient procedures for reducing internal costs, new services, etc. Often a new company is founded mainly as an embodiment of a new business idea (e.g., setting up a pizzeria in a part of the town where there were previously no restaurants; selling dog food on the Internet). Business innovation is the ordinary working of competitive markets. Patents are an exception to competition rules, which is welfare improving in particular fields where pure competition would erase the reward needed to incentivize invention. Extending patents to all areas of business activity means that all dimensions of competition would be subject to restrictions unjustified in terms of social welfare.

5.3. **Inventive Step**

One basic condition for an invention to be patentable is of course that it should be novel:

An invention shall be considered to be new if it does not form part of the state of the art. The state of the art shall be held to comprise everything made available to the public by means of a written or oral description, by use, or in any other way, before the date of filing of the European patent application.

(EPC Art. 54, paragraphs 1 and 2)

Box 5.4. The problem/solution approach of the EPO (Guidelines Part C, chap. IV, section 9.8: p. IV–24 sq.)

The 'problem-and-solution approach' (PS) is a method developed at EPO between the late 1970s and the early 1980s, with the view to establish routines to assess the inventive step of an invention, so as to harmonize the decisions coming from examination (a crucial issue at the time the EPO was created, as it recruited examiners coming from various national offices with totally different approaches to examination, carrying the risk of inconsistent decisions across the EPO) and remove as much subjectivity as possible from the decision.

The PS approach works in three main stages:

1. Determining the closest prior art, which is defined as 'that combination of features, disclosed in one single reference, which constitutes the most promising starting point for an obvious development leading to the invention . . . . In practice, the closest prior art is generally that which corresponds to a similar use and requires the minimum of structural and functional modifications to arrive at the claimed invention'.

2. Establishing the 'objective technical problem' to be solved: 'To do this one studies the application (or the patent), the closest prior art and the difference (also called "the distinguishing feature(s)" of the invention) in terms of features (either structural or functional) between the invention and the closest prior art and then formulates the technical problem'. The objective technical problem as established by the examiner might differ from the one as viewed by the applicant.

3. 'Considering whether or not the claimed invention, starting from the closest prior art and the objective technical problem, would have been obvious to the skilled person'.

But mere novelty is not enough: the invention should be 'significantly novel' or significantly different from the state of the art—it should involve an 'inventive step'. 'An invention shall be considered as involving an inventive step if, having regard to the state of the art, it is not obvious to a person skilled in the art' (EPC Art. 56).

The US equivalent to inventive step is 'non-obviousness', which has essentially the same contents (although the implementation differs somewhat). Determining whether a given invention reaches the inventive step is not totally exempt from subjectivity. However, there are rules, standards given to examiners, which aim to limit that subjectivity. These rules are a possible leverage for patent policy, to be used when policy makers are willing to strengthen or to weaken conditions for patenting, in order to influence innovation or competition.

In practice, the inventive step is assessed at the EPO by using the 'problem–solution approach' (see Box 5.4.)

The grounds for requiring novelty are clear enough, as granting a monopoly on an existing invention would have no incentive effect while imposing costs on society. But why become stricter and require an inventive step? Three main reasons justify the requirement of an inventive step by the patent office: (1) maintain market competition; (2) reduce uncertainty; and (3) ensure sufficient protection to inventors.

The first reason is about ensuring that a patent on a small incremental invention does not endanger established businesses. As it was made clear by decisions of British courts as early as the seventeenth century (see Chapter 2) routine innovation is part of the ordinary way of conducting business. A company will normally adapt its product to the requirement of particular customers; it will fix a default to its products; etc. Allowing another company to patent such marginal changes would equate to forbidding established companies to conduct their own daily business; requiring only mere novelty would hurt market competition.

The second reason is that a low inventive step is more difficult to implement safely, and with adequate precision, than a high one. As there is some uncertainty regarding the decision of the examiner (i.e., the appraisal of inventivity is necessarily noisy and cannot be perfect), a low inventive step makes it more likely that certain patents that are granted will simply not be novel. They would therefore drive out of business established companies on the grounds that they infringe patents which in fact are not even novel. Taking into account the possibility of a mistake by the patent office or the courts, requiring an inventive step is a way to ensure that simple novelty is implemented.

The third reason refers to the effectiveness of the patent system. A patent system with a too low inventive step cannot guarantee inventors an economically significant reward. In a framework of cumulative invention, an inventive step is necessary for the patent system to work, to incentivize inventions. In the absence of inventive step, any invention could be immediately superseded by a marginal, simple improvement, so that the initial patent would be in fact void. This was actually the case with 'brevets de perfectionnement' in France in the nineteenth century, whose effects are expressed in Balzac's novel, *Les illusions perdues*:

La plaie des inventeurs, en France, est le brevet de perfectionnement. Un homme passe dix ans de sa vie à chercher un secret d' industrie, une machine, une découverte quelconque, il prend un brevet, il se croit maître de sa chose ; il est suivi par un concurrent qui, s'il n'a pas tout prévu, lui perfectionne son invention par une vis, et la lui ôte ainsi des mains.

Modern economics goes further. Models by O'Donoghue (1998) and by Hunt (2004) point to the economic trade-off to which the choice of the inventive step is a solution. A low inventive step will (1) reduce the expected life-time of the patented invention, as improvements superseding it while not infringing its patent will come earlier, hence reducing the expected gain to the inventor; and (2) encourage smaller inventions at the expense of larger ones, as the former can be protected and can supersede the latter while they are less costly to do. In other words, a lower inventive step encourages smaller inventions and creative destruction, but it discourages large ones. Conversely, a high inventive step will (1) lengthen the expected duration of monopoly and foster the reward to patented inventions; and (2) discourage minor improvements, which would be barred from the market or at best would require that licence fees be paid to the incumbent.

Where the balance should be set is not straightforward and certainly depends on technological fields. Small inventions have their utility, as they increase the value of products to customers. Also, certain large inventions come step by step, and preventing the patentability of each step will induce the inventor to keep secrecy on these intermediate steps whereas society would benefit from their disclosure (as it would allow other inventors to enter the inventing race without having to duplicate findings from previous steps).

The legal approach emphasises that the inventive step is purely technical and that it is uniform across technical fields. The reality is more complex. The inventive step is in fact a highly flexible and rich instrument, which can be adapted to the diversity of technical characteristics and to various particular circumstances, and it integrates factors which go beyond mere technical considerations.

Neutrality vis-à-vis the technical field is impossible to check, even to define in operational terms, as there is no metrics of the inventive step: how can one compare the inventiveness of a new molecule with the inventiveness of a new screw? In view of the diverse conditions prevailing across technical fields, only a differentiated standard could guarantee that the result is 'neutral'. At the core of the test is the notion of the 'person skilled in the art', who is defined as an ordinary practitioner aware of what was common general knowledge in the relevant art at the date of invention. He or she also had access to everything in the 'state of the art', in particular the documents cited in the search report, and had at his disposal the normal means and capacity for routine work and experimentation (see Guidelines for examination, Part C, Chap IV, pp.IV–22).

This person is perfectly informed of the knowledge in the relevant art and is totally rationale. He or she is supposed to be largely deprived of creativity, being able to conduct 'routine work and experimentation' only. However, the notion of routine experimentation is clearly different across industries. In industries which are more recent, with abundant technical opportunities, such as information technology or biotechnology, 'routine experimentation' could lead to significant inventions (in terms for instance of product improvement). Conversely, in mature technology fields such as chemistry or mechanics, in which the productivity of research (roughly defined as the impact of a given amount of effort on the performance of the improved product or process) is lower, significant efforts could result in minor improvements. The person skilled in the art integrates that diversity: he or she is, as the name indicates, adapted to the particular art in which he or she is active, and which differs from other arts. One expects therefore the inventive step in each field to reflect that specificity.

Other criteria are used as well that vary the requirements for patentability across technical fields. For instance patent applications for genetic inventions should document 'substantial and specific functions' (although the protection extends beyond that—it is a product patent, which includes the genetic code). There is no such requirement for chemical compounds, making the inventive step de facto lower in this field. This is consistent with recommendations from economic analysis (Hunt 2004). The inventive step should be higher in fast evolving fields than in slow moving ones. The rationale for the inventive step being to ensure sufficient life time for existing patents, the faster the technology evolves the shorter is the expected life time of patents for a given level of inventive step (as they will be superseded by superior technology). Therefore in order to equate the expected lifetime across technical fields so that the incentive effect of a patent would be similar in biotechnology and in chemicals for instance, one should adapt the inventive step requirement accordingly.

When looking at EPO case law, one can identify a series of features that are concordant with the economic approach:

- Commercial success: the fact that a field is demonstrated to be 'economically significant and frequently studied' can be used to argue for inventiveness of an invention made in that field, on the ground that despite much effort nobody found it before (EPO 2001: 135). Market success can be used to demonstrate an inventive step, but only as 'secondary indicia' (it must be shown that the commercial success was due to the technical features of the product, not to marketing). The fact that there is commercial success is interpreted as reflecting the existence of a significant need for that invention. As persons skilled in the art are supposed to know their market, if they had been able to solve the technical problem corresponding to this new invention, they would certainly have done it already. The fact that they did not do it demonstrates its non-obviousness.

- Small improvements at large scale: 'Even small improvements in yield or other industrial characteristics could mean a very relevant improvement in large-scale production' (EPO 2001: 131). Or 'Since a process of this kind was obviously intended for the production of paper on an industrial scale, even a small improvement had to be regarded as significant' (EPO 2001: 131) So, a given improvement in technology (say, X percent increase in the efficiency of a process) will be deemed patentable if it can be implemented at large scale, but not if it concerns only small ranges of production. As it is made clear in the second quotation, this criterion is applied at the individual industry level. The reasoning here, as above, is that large scale indicates high value, and if it had been easy to invent that would have been done already in view of so high value.

- Could–would: if the invention was realized by combining different components, obviousness would not be proved by the fact that the persons skilled in the art *could* have assembled the various components, but by the fact that they *would* have done it, meaning they would have had good reasons to do it–including economic reasons (e.g., demand, competition, etc.).

The point is not whether the skilled person could have arrived at the invention by adapting or modifying the closest prior art, but whether he would have done so because the prior art incited him to do so in the hope of solving the objective technical problem or in expectation of some improvement or advantage.

(Examiners guidelines, C, IV, 9.8.3)

It is a routine (not inventive) activity for the persons skilled in the art to just follow prevailing trends in the industry:

Mere automation of functions previously performed by human operators was in line with the general trend in technology and thus could not be considered inventive. The mere idea of executing process steps automatically, e.g. replacing manual operation by automatic operation, was a normal aim of the skilled person.

(EPO 2001: 133)

As these trends are dictated not by technology only but also by economic factors (orientation of demand indicating what consumers are ready to pay for, cost structure of the production process indicating which components of cost should be reduced first, etc.), a patent will be granted only if the invention was also non-obvious in technical terms *in view of the economic incentives facing the inventor.*

The flexibility of the requirement for inventive step is illustrated by the existence of particular types of invention protection which do not require the same inventive step as patents, notably 'utility models'. Utility models, or 'petty patents', are types of patents which provide a weaker protection (usually ten years instead of 20 for standard patents), for a lower cost and with lower requirements. Utility models exist in many countries, including in Europe. Usually utility models are granted without any prior examination: in certain countries examination would happen only if requested by a third party (in an attempt to revoke the title) or in case of litigation. The requirement for inventiveness, labelled 'innovative step' in Australia for instance, is somewhere between simple novelty and a mere inventive step. In certain countries such as Germany or Austria, the law authorizes inventors to file for a patent and a utility model at the same time for a given invention. That results in applicants having a fall back position in case their invention is not inventive enough for getting a patent, but is still significantly novel.

The requirement for inventive step can also be adapted to policy considerations. There is in the EPC (but not in US law) a special provision that allows the inventive step to be adapted in certain particular circumstances (EPC Art. 54(3) 56). That happens when two patents are filed by two different entities for similar inventions, the second one being filed before the first one has been published. This is a case of quasi tie in a patent race. Applying strictly the general rule would lead to reject the second application, on the ground of non-novelty or, more likely (as two inventions are rarely perfect duplicates) lack of inventive step. In that case, the EPC states that the first patent will be considered as prior art in the search for novelty of the second one, but it will be excluded for the examination of inventive step (otherwise the chances for the second patent application to be granted are close to zero). In other words, the loser in a patent race could still have a patent granted, but it will be a narrower patent. That approach allows softening the 'winner take all' feature of the patent system. It is not uncommon in certain areas such as biotechnology or chemistry, fields where it is estimated that about 5 percent of grants are made under such conditions.

5.4. **Scope**

The scope, also called the breadth, of a patent defines the technical domain which is protected by that patent, i.e., the set of inventions which would infringe that patent. The scope of a patent has direct economic effects: it increases the private value of the patent and its cost to society, including reduced value of patents covering competing inventions. Broader scope has two distinct effects: (1) it makes it more difficult for competitors to enter the market and offer substitutable products; and (2) it makes it more difficult for other inventors to develop further inventions based on the initial one. These two effects are addressed separately in the economic literature, under the names of 'lagging breadth' and 'leading breadth', respectively (Scotchmer 2004). If the invention is integrated into a final product (e.g., a new toothpaste), the breadth of the patent will affect horizontal competition, it will determine how far competitors should position their own products in order not to infringe—this is the lagging breadth; if the invention is an upstream one, e.g., a new genetic pathway that is used in further research, the breadth of the patent will affect more the incentives to do downstream research using the initial invention—this is the leading breadth. Certain inventions, for instance drugs, are both sold to customers and could be used for further research (for possible second use).

The issue of scope is well illustrated by the 'oncomouse patent', a patent on a genetically modified mouse especially fitted to study cancer, held by Harvard University. The patent as initially granted for the oncomouse actually covered all genetically modified mammals (in 1988 in the US, in 1992 by EPO, where humans were excluded from the scope in view of the EPC—Publication number 0169672). It was then subject to an opposition procedure at the EPO and narrowed down to rodents in 2001. The decision was appealed and the patent was finally reduced to mice only in July 2004 (Decision T 0315/03). The exclusion power of a patent covering mice is obviously much narrower than that of a patent covering all mammals. The issue of scope is also clearly raised in the pharmaceutical field: should the second use of a drug be included in the scope of the patent protecting this drug, although this second use was not foreseen by the drug's inventor, or should a further patent be granted to the inventor of that second use?

The legal approach to patent scope addresses the technical properties of the invention itself, trying to identify the boundaries of what the inventor actually found or not. This approach works well in mature technical fields, where technology is quite stabilized and well defined. In emerging fields, where radically new inventions happen frequently, technological breakthroughs, redefinitions of the field, new perspectives, etc. an inventor will often have both real findings and suspicions, or intuitions, regarding further implications and applications of his or her invention and future ways for the technique to develop. These suspicions can have much value as they will trigger future research, but they might not be as 'hard' and proven as real findings. The boundary between a real finding and a suspected possible extension is blurred, as extensions might rely on interpretations which are shared by some scientists but not all. In these fields closely tied with science, there could also be doubts regarding the boundary between the invention per se, which is patentable, and its natural background (e.g. a material or a mathematical formula), whose patentability can be excluded by law. In all these cases where no clear answer is provided by purely technical considerations, an economic approach would allow adequate legal rules to be founded. The economic approach balances costs and benefits of expanding or restricting the scope of a patent, in terms of its impact on the incentive to innovate, deadweight loss and the diffusion of knowledge for future research.

THE LEGAL APPROACH

The scope of a patent is defined by the claims: 'The claims shall define the matter for which protection is sought. They shall be clear and concise and be supported by the description' (Art. 84 EPC).

For patent offices, scope is closely related to inventive step. The basic rule is that what is documented in the specifications, and only that, can be claimed, provided that it involves an inventive step (and satisfies other conditions of the EPC). This approach is explained in the EPO Board of Appeal Case Law book (EPO 2001: 101):

The extent of the monopoly conferred by a patent should correspond to and be justified by the technical contribution to the art. This general principle of law [applies] to determine the scope of protection justified under Art. 83 EPC and Art. 84 EPC, also applies to decisions under Art. 56 EPC, because everything covered by a legally valid claim has to be inventive. Otherwise the claim has to be amended, by deleting anything obvious to ensure that the monopoly is justified.

Hence, a low inventive step will lead to broader lagging breadth (more aspects of a given invention will be assessed as being inventive).

The inventive step influences scope also in an indirect way. It determines which other inventions, based on the initial one, are patentable or not. A low inventive step will result in a lower de facto leading breadth, as minor modifications of the initial inventions will be patentable. A new chemical compound differing only slightly from an existing one (and possibly copied from it), having similar properties, would be patentable and put its owner in a position to seize the market.

The scope of a patent is revealed when a court examines a case of alleged infringement—asking whether another product is at least partially covered by the patent. Article 69 EPC applies to infringement proceedings in the domestic courts. In addition, the European 'Protocol on the interpretation of Article 69' (also signed by EPC member states) states that the claims, when ambiguous, should be interpreted in the light of the specifications: 'The extent of the protection conferred by a European patent or a European patent application shall be determined by the terms of the claims. Nevertheless, the description and drawings shall be used to interpret the claims'.

Despite this common basis, two opposite doctrines are referred to by the European courts: the doctrine of equivalents and the doctrine of literal interpretation. 'The effect of the doctrine [of equivalents] is to extend protection to something outside the claims which performs substantially the same function in substantially the same way to obtain the same result' (House of Lords 2004). Conversely, the doctrine of literal interpretation states that protection is conferred only to what is claimed explicitly in the patent document. It proscribes explicitly 'anticipatory claims' (claims that cover possible future extensions of the invention, yet to invent) and 'insufficient disclosure' (certain claims are not supported by corresponding specifications). The 'Protocol on the interpretation of Article 69' appears to be a compromise between the two approaches.

British courts explicitly rule out the doctrine of equivalents (House of Lords 2004), keeping to a restrictive interpretation of Article 69. On the other hand, under German case law modifications of a patented invention by a third party could also be found as infringing under the doctrine of equivalents.

A modification still falls under the scope of a patent if: (1) the modified means have objectively the same effect as the means of the patent; (2) the person skilled in the art was able to find such modified means; (3) the person skilled in the art would also consider such modified means as a solution of equal quality to what is patented when reading the patent.

(EPO 2005: 56)

Hence scope is one case where the practice differs across countries within Europe, making it possible that a patent be revoked or narrowed down in one country while being maintained in its initial setting in another country.

The scope of patents is also influenced by other factors, such as exemptions which limit the exclusion right conferred to the holder (notably for research use, see Chapter 3), or special features of patent law such as 'product by process claims' that concern essentially chemistry (Art. 64(2) EPC: 'If the subject-matter of the European patent is a process, the protection conferred by the patent shall extend to the products directly obtained by such process').

ECONOMICS: HORIZONTAL COMPETITION

The Watt-Boulton patent of 1769, protecting James Watt's seminal invention, was extremely broad, covering 'a method of lessening the consumption of steam and fuel in fire engines'. It left no room for inventing around. Armed with such a patent, Watt could keep his monopoly over the steam engine industry for about 30 years, as he defeated all challenges in the courts. Watt also refused to grant licences. Hence he established an absolute monopoly on the market. When the patent lapsed, in 1800, in only a few months' time there was a blossoming of improvements put on the market by competitors, which had been waiting for years in the workshops (Daumas 1964).

In a static, competitive framework, the scope of a patent determines the intensity of competition that the patentee will be faced with. Scope can be modelled in different ways. It can be represented as the maximal degree of substitutability between the patented invention and a competing one which is allowed without being deemed to infringing the patent (Klemperer 1990). In that framework, broader scope means that competing products will be less substitutable to the patented one, whose holder could therefore exercise stronger market power. Alternatively, scope can be represented as the cost of inventing a substitutable product that would not infringe on the patent (Tandon 1982). In that case, a broader scope means a higher entry barrier to the market, resulting in a smaller number of competitors and ultimately weaker competition. Both representations show a positive relationship between scope and the mark up over cost that the patentee could charge.

The issue of optimal scope in a static framework is about how profitable a patent should be (Scotchmer 2004). Its determination will result from balancing the incentive effect (more profit is good) and the deadweight loss (low markup is good). Optimal scope is closely related, in the theoretical literature, to duration, as it is the combination of the two which determine the total reward to the inventor: scope determines the reward in each period, while duration tells for how many periods the inventor is entitled to earn it. The problem is then how to structure the total profit between scope and duration so as to ensure sufficient reward to the patent holder while minimizing the market distortion. Broader scope implies less competition, hence a higher markup in each period of time, but it allows a shorter duration of the distortion. Longer duration means that the markup gained in each period of time could be lower. As pointed out by Gilbert and Shapiro (1990), the crucial element in this trade-off is whether breadth increases profits more rapidly than it increases the deadweight loss. If it does, a regime with broad scope over a short period of time is preferable. Otherwise, narrow but long-lived patents are better. As the impact of breadth on profits and deadweight loss is hardly measurable, it is difficult to draw general, operational conclusions from that strand of literature. The deadweight loss depends on factors such as the cost of inventing a substitutable but not infringing product (if it is low, a broad patent will not damage too much competition), substitutability of goods (loss of utility of consumers when they switch from the patented good to the closest one. If it is low, then a broad patent will do little harm). We will come back to these issues when addressing duration of patents at the end of this chapter.

ECONOMICS: SEQUENTIAL INVENTIONS

The issue of optimal leading breadth arises notably in the context of sequential inventions, i.e., when an invention is a direct and necessary step for another one to happen (see Box 5.1). This is exemplified by 'basic patents' on upstream inventions, which open new avenues of research. Such situations arise in the field of biotechnology, where upstream inventions (e.g., genetic material, stem cells) are patented while having no direct utility but serving as a starting point to invent medical treatment. The issue raised is: to what extent should the basic patent pre-empt protection on later developments, or, in the words of a US judge, 'pre-empt the future'?

A situation of 'sequential innovations' arises when the final product delivered to customers results from several successive stages of research conducted by separate entities (see Box 5.1). The revenue is generated at the last stage, when customers pay for the final product, but has then to be shared between the various contributors. Let us restrict the analysis to a simple, two-stage case, which is more tractable analytically and corresponds to most industrial situations, hence separating basic research performed at the first stage from applied research done at the second stage. The leading breadth of a patent at the first stage is defined as follow up quality improvements that would fall under the patent. Leading breadth should be addressed from two angles: the incentive to do the basic research vs. the follow up research (sharing the reward); and the disclosure of basic results so as to allow other parties than the initial inventor to participate in the second stage.

The issue for society (represented by the patent office or the court) is as follows: how should the final revenue be shared between the two stages of the process so that both are performed and that the total cost of research is minimized? Revenue sharing will depend on the scope of patents granted at each stage. One example is a drug resulting from the analysis of genetic material: its invention would require the identification of the gene, of its function, of its involvement in a particular disease, and the invention of a cure that would target genetic pathways related to that gene. Patents can be taken at various stages of the process, usually two: at the upstream stage (the gene plus one function); and at the downstream stage (a cure based on that gene and function). Granting too narrow a patent at one stage will reduce the negotiating power of the inventors at that stage vis-à-vis inventors of the other stage, lowering their reward, hence their incentive to invent in the first place, and possibly breaking the innovative chain. Conversely, too broad a patent might lead inventors to require too high a share of the total reward, at the expense of other inventors, who might then choose to keep out, also breaking the chain.

The scope of upstream patents also affects early disclosure of basic inventions (Matutes et al. 1996). In the case of genetic inventions for instance, narrower patents (covering the applications only) could encourage inventors not to publicise the genetic material they have found, or to publicise it only late, when they have worked out and patented as many applications as they could. This would be detrimental to society, as disclosure allows other researchers to investigate further applications of the invention, while secrecy forbids or postpones that possibility. This is the situation currently in protein research, as the conditions of patentability of proteins (notably of their 3D structure) have not been clearly established yet, hence creating a situation of legal uncertainty that leads inventors to take a cautious strategy based on secrecy.

A complementary view is Kitch's 'prospective theory' of patents. Kitch (1977) views scope as a problem of coordination among different researchers working on related technologies. Without coordination, there is likely to be wasteful duplication of effort and over-investment as competing firms try to be the first to break through. Granting broad patent rights to the pioneering inventor as a technology initially develops will rationalize the development process as the patentee will be in a position to control downstream developments, to allocate the various directions of research to different entities including himself. In that process, patents allow the implementation of a kind of centrally planned process. The pioneering inventor has an incentive to include in the downstream development other potential inventors with additional ideas or capabilities, via licensing or other contractual arrangements. Kitch's arguments are still debatable. First, it is not over-investment but under-investment which seems to be the real problem in most areas of research. Second, even if over-investment was the problem, broad upstream patents would not solve it, but move it instead, from the downstream to the upstream stage (as more profits would be realized at that stage, that would attract more investment there). Third, Kitch's argument does not solve the problem that in many cases too broad upstream patents will reduce profitability of downstream research, with the danger that too little of it would be done (hence reducing in turn the profitability of upstream research, except if the patent holders do it themselves).

In the field of biotechnology, genes and proteins can be protected with patents. Such protection is granted only if a specific and substantial function is disclosed for the genetic material, but the patent is still a product patent. It applies to the genetic material itself if the knowledge of this material is new, so that any user of that material, for doing downstream research, should seek agreement of the initial inventor. That comes from the approach adopted for chemical compounds, after decades of controversy in Europe, which acknowledged that such compounds, when isolated from their environment and synthezised by man, would constitute an invention, not a mere discovery.

The question of whether patent protection for chemical compounds should be absolute—covering all uses and functions of the product—or function-limited, in other words limited to the uses and functions disclosed by the patent proprietor, was debated for decades, before absolute product protection finally prevailed throughout Europe. The main reason advanced for awarding absolute product protection for chemical substances is that it is prohibitively expensive to put a pharmaceutical compound on the market due to the huge cost of toxicity testing and clinical trials. It is argued that the proprietors of patents for further medical uses benefit from the work done by the first patent proprietor, so the latter is entitled to a portion of the rewards obtained by the former.

(Yeats 2005; see also Rutz and Yeats 2004)

An opposite approach has been taken with software. Many software inventions consist in implementing a mathematical formula in a specific way or for a specific purpose. In that field, a patent covers only the specific way or purpose of implementing the formula, while the formula itself remains in the public domain. Such patents are much narrower than those granted in the field of biotechnology.

DEPENDENT PATENTS

Standards applied by patent offices can be such that the leading breadth is higher than the inventive step in certain cases, for instance with genetic material, chemical compounds, and drugs. That means that certain inventions are considered to be inventive in the sense of patent law, but fall within the scope of an already granted patent. In that case, these new inventions can be granted a 'dependent patent', which is a normal patent but whose implementation requires a licence from the holder of the first patent. 'In cases where the holder of the main patent cannot exploit its invention either without harming the rights of the holder of the dependent patent, the majority of laws in EU allow for the granting of crossed or reciprocal compulsory licences' (IPR Help Desk 2004). Dependent patents could be obtained mainly for new processes for producing a known and patented product, or for new uses of that product.

A case in point is 'second (further) medical uses'. It took a decision of the Enlarged Board of Appeal for having such patents accepted by the EPO, under certain conditions (G 1/83). The legal difficulty is that a further use for a known substance could be simply considered as a therapeutic method, which is not patentable. That difficulty was solved by a recommendation to the applicants to draft 'Swiss-type claims', in the form 'use of a substance or composition X for the manufacture of a medicament for therapeutic application Z' (EPO 2005a). The key-word, legally, is 'manufacture'. Such patents could be granted to new uses found for an existing drug (diseases to be treated not foreseen by the original patent holder), for 'the application of a known medicament for the prophylactic treatment of the same disease in an immunologically different population of animals of the same species' (EPO 2001: 91), for a new mode of administration of a known drug, etc.

A new prescribed regimen of a known drug, for instance, used not to be patentable (decision T 317/95), but was allowed by a decision taken in 2005 (T 03/1020, 'allowing Swiss form claims directed to the use of a composition for manufacture of a medicament for a specified new and inventive therapeutic application, where the novelty of the application might lie only in the dose to be used or the manner of application' (paragraph 72), justified on economic grounds:

A change to this narrow focus view [which the Board did not take] on novelty would not be in the interests of those doing research in pharmaceuticals. It would probably slow the increase of knowledge on how medicines can most effectively be used. It might cause a reduction of prices, but not even this would be certain if there was simply a shift of promotion and research expenditure to products for which patent protection was obtainable.

(EPO 2005: paragraph 75)

Dependent patents are also used sometimes by original patent holders as complements to their own initial patent: In order to broaden its scope or to extend its duration beyond the statutory term (the new patent overlaps with the initial one, hence forbidding its use to third parties). This practice, along with others having the same aim, seems to be used by certain pharmaceutical companies with the purpose of delaying the entry of generic drugs in markets.

EMPIRICAL STUDIES

Empirical studies on the effect of the scope of patents on inventions are scarce, due notably to the difficulty to measure scope. The measure most often used is the number of claims. That is not straightforward as in many cases claims are added to patent applications in order to limit their scope (disclaimer), not to broaden it. But apparently the number of claims is well correlated with other measures of the size of the protected invention, such as the number of citations (forward and backward) or the number of members of the family (Lanjouw and Schankerman 1999). For studying the impact of breadth on innovation, Sakakibara and Branstetter (2001) investigate the case of Japan, where a reform introduced in 1988 allowed patents with multiple claims (whereas the number of claims was limited to one before). Based on interviews with industrialists and statistical investigation, they find only a small positive effect of increasing patent breadth on R&D expenditure of a panel of Japanese companies.

5.5. **Duration**

You own a piece of land forever, but you own a patent for a limited duration only. The statutory life of patents is nearly uniform worldwide: the TRIPs requires signatory countries to fix it at 20 years minimum (from the time of application), which is exactly the number chosen by almost all countries. It used to be 17 years, either from application (Germany) or from grant (US) in a number of countries. The EPC had already brought harmonization in the duration of patent life across Europe in the 1970s.

Limited duration is one major difference between property over tangible goods and intellectual property (except trademarks). Why a limited duration? That feature of intellectual property rights reflects the mission they are assigned by society. They are a policy instrument, whose validity should stop at the point where the costs to society start exceeding the benefits.

The reward provided by a patent should compensate the inventor for the cost and risk of that the invention required. The total reward provided by a patent is the sum of rewards per period, cumulated over its life time. Hence the total reward is increasing with the duration of the patent. Under current law (20 years), and assuming a constant flow of revenue over time and a discount rate of 10 percent per year, a simple actuarial calculation shows that a patent holder appropriates 86 percent of the discounted value generated by the invention if it was protected over an infinite term. Hence, in economic terms the 20 years term is not so short, as the inventor appropriates the bulk of the value of the invention (for copyright, whose term is 90 years in the US, the figure is 99.997 percent). Hence 20 years is already quite generous to the average inventor. It might be less the case for seminal inventions and to a certain extent for drugs, which see their value increasing over time.

Another reason for limiting the term of patents relates to transaction costs. Inventions do not survive unchanged over time, they mix with each other, they are transformed. Hence any single contribution to new technology gets progressively diluted in the flood of subsequent inventions. Keeping track of any single inventor's own contribution therefore gets difficult over time, until the point when it is simply impossible. Imagine the situation if society were should still reward Louis Pasteur's, Thomas Edison's or Rudolph Diesel's heirs today. That would create conditions for permanent litigation between the heirs and heiresses of great inventors.

A finite life for patents is justified: but how long should it be? The economic literature on patent design started with that question, addressed in a seminal study by Nordhaus (1969). The optimal duration is obtained by minimizing the discounted value of the deadweight loss generated by the patent over its entire life under the constraint that the discounted profit exceeds the cost of the invention (adjusted for risk). That pleads for a differentiated duration, which would depend on characteristics such as the cost of the invention and the shape of the deadweight loss as a function of price and time.

Subsequent developments in the economic literature have linked duration with scope. As shown above, the same revenue could be obtained by the inventor either by changing the scope or by changing the duration of the patent. The pair (i.e., the breadth and duration) which is optimal for society depends upon the impact of the price on the deadweight loss. Long and narrow patents will be preferred if the impact of price on the deadweight loss is convex, in other words, if the dead weight loss increases more than proportionally with the price of the product (showing that the price elasticity of the product is high) (Tandon 1982; Gilbert and Shapiro 1990). Broad and short patents are preferable in the opposite case. Hence, using different assumptions, different authors end up with opposite conclusions: Gilbert and Shapiro favour long and narrow patents, while Gallini (1992) advocates short and large patents. The discount rate also plays a role: the higher it is, the more duration should be expanded for compensating a reduction in breadth as the discount associated to future revenues is higher.

On the ground that the duration of a patent should be smaller when cost and risk of the invention are lower, it has been proposed by Jeff Bezos, CEO of Amazon, that patents on business methods be restricted to six years.

The trade-off between scope and duration is also affected by risk, although in an asymmetric way. A broader patent brings higher profits, but it also brings more risks, so that patentees will not systematically seek the broadest patents from the patent office. A broader patent is more likely to be found infringing another patent by a court (validity is less predictable with a broad patent than with a narrow one). On the other hand, longer duration brings its own risks, as many unforeseen events could happen in the future. The invention could be superseded by a superior substitute not yet invented; consumer's demand could vanish, etc., so that the patent would loose economic value before its statutory term ended. In situations of cumulative inventions, patents are abandoned when the invention is superseded by a superior one. If minor improvements over an invention can be patented, that invention will loose its own economic value. A patented invention will be overtaken more rapidly the smaller the leading breadth is, and the smaller the required inventive step is. Hence leading breadth, inventive step, and effective life of patents are connected (O'Donnoghue et al. 1998).

The actual duration of a patent is not necessarily its statutory life. In order to keep their rights valid, patent holders must pay renewal fees to the patent office in most jurisdictions. Such fees are to be paid on a yearly basis in most European countries, every three and a half years in the US (the last renewal fee being due in year 11.5, which allows keeping the patent in force up the twentieth year). Payment of renewal fees to the EPO starts at the application stage (annual fees are due from the third year after application) and stops at grant, as the patent is then transferred to national patent offices, where national renewal fees apply. If the patent holder does not pay the fees the patent lapses, falling into the public domain. The decision not to pay is irreversible (i.e., a dead patent cannot be revived). Renewal fees lead patent holders to self-select inventions they want to keep protected. The holders will maintain the patent in life if they estimate that the expected discounted income this patent will generate is higher than the discounted cost of maintaining it. Hence, renewal fees are a device that allows the patent holder to choose the duration of the patent on the basis of the expected private value of the patent and its time structure. In fact, 50 percent of EPO patent lapse in the first ten years after filing, while about 8 percent only are renewed until the statutory term (most of them are related to pharmaceutical products).

Scotchmer (1999) defines a patent mechanism for an isolated invention by two parameters: the price to be paid to obtain protection (lump sum patent fee) and the corresponding length of protection acquired (patent length). The patent mechanism is described as a menu of different pairs of patent fees and lengths, where fees are a non-decreasing function of length. Each applicant chooses a pair from this menu according to the expected monopoly profit per period. Thus, innovators face a trade-off between either paying a higher patent fee to obtain a longer patent life and capturing a higher overall profit, or paying a smaller patent fee (or even receiving a subsidy from the public authority in some cases) for a shorter patent life protection. Inventors with high value inventions may want to pay for longer protection, even if benefiting from a longer protection is made more costly (see also Encaoua et al. 2004). Hence renewal fees have the virtuous property of being a 'revelation mechanism' in the sense that the choice made by the patent holder reveals information that society can use for orienting the holder's choice in accordance with the interest of society. If private value is an increasing function of market exclusivity, and if social value is increasing in private value, then renewal fees result in shortening the market exclusivity of inventions with lower social value.

A fee-based mechanism is not perfect for selecting patents. First it uses the private value as a proxy for the social value. The social and private value are not necessarily in the same proportion of each other for all inventions, it can even happen that the private value exceeds the social value: then, basing selection on the private value might lead to excluding inventions whose social value, even not high, exceeds the private value, while it would accept inventions with high private value even if the social value is low. Second, the supplement in fee adds to the cost incurred by the inventor and will (1) tend to deter research (of which the cost in increased); and (2) will be partly passed to customers, hence increasing the deadweight loss.

A feature common to a large number of countries is that renewal fees are rais-ing over the life time of the patent. For instance, in Germany the progression is steady, from about €100 in the fourth year to €2000 in the twentieth (2005 fee schedule); in France the progression is step by step, from about €100 per year until the tenth year, to more than €500 beyond the sixteenth year (2005 fee schedule). Switzerland is a rare exception in that regard, as it abandoned progressivity in the late 1990s for adopting a constant renewal fees system.

Why are renewal fees progressive? A patent which is renewed over a longer period of time is presumably of higher value. An increasing renewal fee is then a way for government to extract more revenue from wealthier inventors. Actually, Cornelli and Schankerman (1999) argue that, in view of the higher value of patents whose life is extended the most, renewal fees are in fact regress-ive taxes—they increase less than the value of the underlying patents. Higher taxation on the wealthiest ones allows lower taxation of others: it is a way of lowering the entry cost to the system while preserving the financial balance of the patent system (covering administrative costs), as most of the revenue comes from successful inventors. This is justified when the government wants to attract more patent applications, including from inventors uncertain at the early stages of the future value of their invention. Conversely, upfront fees should be set higher if the issue is to reduce the number of applications.

The statutory duration of patents is in principle uniform across technical fields: However, in rare cases it is also fitted to industry specificities. This hap-pens with drugs. As it takes ten to 12 years to have a drug marketed after the patent was filed, that leaves eight to ten years for the drug to generate revenue to the company. These ten to 12 years are devoted to further research, clinical trials, and obtaining the authorization to market the drug from health administration. Based on the view that eight to ten years of monopoly is not enough for recover-ing the cost of doing R&D, and that part of the delay is due to the government (administrative procedures), the US congress passed in 1984 the Hatch–Waxman Act, which extended the life of patents for drugs. This Act also com-pensates pharmaceutical companies for another reform, the so-called Bolar exemption (after its name in Canada), which allows competitors to start devel-oping generic products before the patent expires, so that they can market them the day the drug falls into the public domain. This type of research exemption did not exist before the Hatch–Waxman Act, and as it takes several years for developing a generic copy of an existing drug, it meant that the effective life of patents on drugs was longer than the statutory life (by an average of two years). There is no Bolar-like exemption in Europe in 2005 (except in particular coun-tries) but this research exemption is quite broad in many countries.

Legislation similar to Hatch–Waxman was passed in Europe in 1992 (Regulation 1768/92), creating 'supplementary protection certificates' (SPC). The term of an SPC is up to five years. Its exact duration depends on the date of issue of the first marketing authorization for the drug in Europe. SPC extend the effective life of patents for drugs to about 15 years after commercialization. The European Parliament adopted in September 2005 a further regulation which offers drug companies an additional six months patent protection on a medicine if they commit to test its effects on children (drugs makers wanted a 12 months extension). In addition, there have been discussions at the US Congress of extending the patent term to biodefense drugs and vaccines, while proposals were made in the UK to extend such a scheme to the development of new antibiotics. These examples illustrate the fact that government already use patent law in a somewhat discriminatory way, in order to reach particular policy objectives. This is beyond the 'one size fits all' principle.

5.6. **Summary**

- 'Patent design' is the discipline which studies the substantive features of the patent system (i.e., the subject matter, inventive step, scope, duration) from the point of view of their economic effects.

- From an economic perspective, a 'good patent system' is one which (1) encourages inventions; and (2) does not hamper diffusion. This double test is applied when evaluating any particular feature of a patent system from the point of view of 'economic quality'. For instance, no patent should be granted to an invention when this invention could be done and publicised without a patent.

- The patent subject matter is fixed by reference to the notions of 'industrial application' (EPC) and of 'technicality' (EPO case law). The list of exclusions includes scientific theories, aesthetic creations, etc. It does not rely explicitly on economic reasoning. Recent extensions of the subject matter have encompassed biotechnology, software (although not 'as such'), while Europe rejects business methods patents.

- The inventive step should be high enough for protecting patent holders from marginal improvers and for preserving market competition. Accordingly, it should be higher in the most dynamic technical fields.

- The scope determines, together with the duration, the income that the inventor will make from the patent. It should be sufficient to reward the inventor, taking into account risk, while minimizing the deadweight loss of customers.

In the case of sequential inventions, the scope determines the sharing of revenue between upstream and downstream inventors.

- The statutory duration is fixed at 20 years in the EPC, in accordance with the TRIPs. Renewal fees can be used for leading patent holders to choose the duration that minimizes the deadweight loss while compensating them for their cost and risk.

Part 2

The European Patent System

6 Patenting Procedures and Filing Strategies at the EPO

Niels Stevnsborg and Bruno van Pottelsberghe

In this chapter we present the EPO patenting process, and investigate what filing strategies assignees can adopt to suit their broad innovation and IP strategy. The first section describes the various routes that are available to file a patent at the EPO. The following section provides an in-depth investigation into the drafting styles that can be adopted to write a patent application. Section three presents the various stages between a patent application at the EPO and its grant, including potential oppositions. The granting procedure takes time, and the behaviour of the applicant regarding the interaction with the office might increase or reduce the period of time between the filing of an application and grant. The last section summarizes the patent filing strategies and the way they could fit a broad patenting strategy.

6.1. The Routes to the EPO

The EPO was established by the European Patent Convention (EPC), signed in Munich in 1973 and entered into force in 1977. The EPO is the patent granting authority of Europe, and its raison d'être is to examine patents. It is the outcome of the European countries' collective political determination to establish a uniform patent system in Europe. The EPO is the executive arm of the European Patent Organisation, an intergovernmental body set up under the EPC, whose members are the EPC contracting states. The activities of the EPO are supervised by the Organisation's Administrative Council, composed of delegates of the EPC Member States, who are often the heads of the national patent offices. The EPO basically examines patents on behalf of all contracting states on the basis of a centralized procedure. The EPO is nowadays composed of 31 contracting states (for the list of membership, and years of entry, see Chapter 2) and five extension states (AL, BA, HR, MK, YU) to which European patents can be extended at the applicant's request.

By filing a single application in one of the three official languages (English, French, and German) and by a single patent grant procedure it is possible to obtain patent protection in several or all of the EPC contracting states, with the EPC establishing standard rules governing the treatment of patents granted by this procedure. The European Patent System becomes complex and costly just after the grant, as a patent must then be validated, put in force, and renewed in each national patent system, with its own legislation and its own fees structure.

The lack of direct Europe-wide market reach is illustrated in Table 6.1. It shows the percentage of the patents applied at the EPO that designated the 13 most frequently designated states, at examination stage. There is a clear preference for the three largest European countries (in terms of population size), as more than 90 percent of the patents designate Germany, the UK, and France. Italy and Spain respectively attract nearly 80 percent and 70 percent of the patents. Smaller countries are designated for protection by 60–70 percent of the patents.

Most often applicants start with a national filing, commonly called a priority filing. The priority filing is important as it provides the first legal date of the application for a patent. According to the Paris convention of 1883, applicants have one year from the priority date to extend their application to other countries or regions. Since the Patent Cooperation Treaty (PCT, signed in Washington, DC on 19 June 1970; and its entry into force occurred on 21 January 1978), the applicant may also file an international application at the World Intellectual Property Office (WIPO), which basically provides more time to decide whether to file an application in foreign countries. After this

Table 6.1. Frequency of designation of EPC member countries at examination stage (2003)

Country	Frequency of designation (%)
Germany	98
UK	94
France	94
Italy	78
Spain	67
The Netherlands	66
Sweden	64
Switzerland	63
Belgium	62
Austria	62
Ireland	61
Finland	61
Denmark	61

Source: EPO Annual Report (2003).

period of 30 months the applicant loses the possibility to obtain a patent in all the designated countries.

Partly as a function of a consideration of expected profits and targeted markets, applicants must take the decision of whether to file nationally, regionally, and/or internationally. Patent applications at the EPO are mainly second filings from national priority applications of EPC or non-EPC member states or PCT applications. Not all the domestic patent filings in Europe are transferred to the EPO.

If applicants want to operate exclusively in their home market, they will probably make a national filing only. This happens when the invention is potentially useful in a narrow geographical area or if insufficient funds are available for a broader geographical scope for protection. With increased financial means and a more international area of operations, filings are also made in other countries, possibly via a centralized procedure such as the EPO.

International filings can then be made by choosing also to file directly at other major patent offices, such as the US Patent and Trademark Office (USPTO) or the Japan Patent Office (JPO), with the option of using the PCT route, having designated these states in the filing documents. In some cases the applicant of an EPC member state might even prefer to file a priority application at the USPTO, and then file either a PCT application designating the EPO or directly file the application at the EPO (Euro-direct applications). Such strategy to file first in the US for instance may be driven by the will to be protected in the US first (for exports); to file through the less expensive and faster process at the USPTO; or for technological reasons (e.g., it is indeed easier to get software of business methods inventions patented in the US).

There are a number of routes that can be chosen to reach the EPO, as illustrated in Figure 6.1. Whereas the Paris Convention provides one year for any priority filing to be extended abroad in any other patent office, the PCT filings allows for a nearly three year (30 months) wait to effectively extend one's application internationally. It is important to keep in mind that in Figure 6.1. any cell of the column 'priority filings' can be followed by any other cell of the column 'subsequent filings'. Most applications filed at the EPO nowadays are second filings following a national priority filing or a PCT filing. For instance, the patent applications at the EPO originating from the United States have generally a priority filing at the USPTO and a subsequent PCT filing at the WIPO. Another important implication is that it is possible to be protected in several EPC member states without going through the EPO. For instance, a Japanese inventor might file a priority application at home, and then file a PCT application at WIPO with a subsequent application at the UK Patent Office (UKPTO) or the German Patent Office (DPMA).

Applicants filing through the PCT route can elect which International Authority shall execute the search during the Chapter I phase or any international preliminary examination, the Chapter II phase. A filing via the PCT

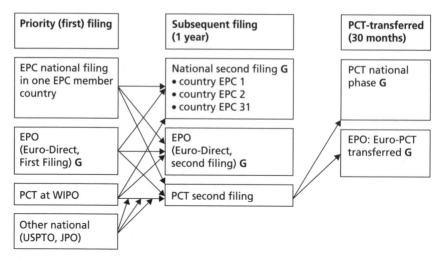

Figure 6.1. Patenting route for protection in one or several EPC member states

Note: The symbol 'G' represents the steps where a granting decision can be taken. Once a patent is granted by the EPO, it must still be translated and validated in each desired national patent office.

Chapter I route, with or without a Chapter II examination, can then be followed up by entry into the European Regional phase before the EPO. There is no requirement that the Chapter I and Chapter II procedures are carried out by the same international search authority. The patent offices which currently (as of May 2006) are authorized to act as International Authorities for Search (ISA) and Preliminary Examination (IPEA) on behalf of WIPO are:

- Austrian Patent Office
- Australian Patent Office
- Canadian Intellectual Property Office
- European Patent Office
- Federal Service for Intellectual Property, Patents and Trademarks (Russian Federation)
- Japan Patent Office
- Korean Intellectual Property Office
- National Board of Patents and Registration of Finland
- Spanish Patent and Trademark Office
- State Intellectual Property Office of the People's Republic of China (SIPO)
- Swedish Patent and Registration Office
- United States Patent and Trademark Office (USPTO)

In the European patent system a first filing at a national patent office (NPO) and a subsequent filing at the EPO directly or via the PCT route can have the added benefit of giving two chances of a valid, granted patent in that particular contracting state. As long as the national filing is maintained and the corresponding added fees and costs are acceptable, and should it not be possible to obtain a European grant, then there might still be a possibility for getting a national patent granted in the end. However, a first filing with an NPO is most frequently performed in a local patent instance, and will be able to provide a first search report and opinion on patentability within a reasonable time delay, and at relatively low costs. Common for all of the above is that there is always the possibility within the priority year to decide on whether to continue the application and which route to follow.

Three important conclusions may already be drawn from the above description of the patenting process. First, it is possible to bypass the EPO for an effective protection in several European countries, even if the majority of patent filings aiming at being protected in more than three EPC member states actually apply at the EPO. Second, applicants must have a clear patenting strategy beforehand, knowing in which countries they want to be protected, and be ready to support the corresponding expenses (see Chapter 7). Third, the patenting process can be highly complex and reflects a fragmented European market for technology, as a granted patent must be managed at the national level of each EPC member state.

6.2. **The Draft of a Patent Application**

The actual drafting of a patent application depends not only on the innovation for which protection is sought but also on the drafting style adopted by the applicant. Writing a 'good' (good being defined as a patent with the desired scope of protection and with a high grant probability) patent requires specific skills for the wording, the writing of claims, the search for prior art, etc.

REPRESENTATION

For drafting and processing their applications, applicants can rely on the service of an official 'representation to the EPO' or, mainly for large firms, can recruit an accredited representative. Representations are governed by provisions of the EPC found in Article 133 and Article 134. Legal residents in any of the contracting states of the EPC may act on their own behalf in proceedings before the EPO. This may at first seems like an option for reducing costs but it is not necessarily a good one, as writing a patent in the proper way is not innate (see for example, Keyworth 2004). However, there are now software products on the market, mainly in the USA, that are especially designed for individuals to prepare their own patent applications. One example of such software is the PatentPro® Patent Writing Software, released in 2004. An example of a patent application written and applied for by the inventor is EP1053772: 'Random pack mass transfer media'.

It is generally recommended to rely on a professional counsel. However, a representative authorized before the EPO must reside in one of the designated states and have passed the EQE (Article 134 EPC: professional representative). Some professional counsels may fall under the term 'grandfather clause' (see Article 163 EPC: professional representatives during a transitional period), which applies to existing patent attorneys in those states which recently became new members of the EPC, and allows them to practice without having the European Qualifying Examination (EQE) qualification.

The choice of external or in-house attorney also depends on the company strategy and financial considerations (Tietze et al. 2006). One might either rely on an external counsel or on in-house competences, or both. Companies that have an in-house patent department also use external counsel, for example for certain inventions that require a specific knowledge not available within the company, or for handling excess workload. As the patent portfolio of a company grows, or with increasing company size, innovation and financial means, there is a natural tendency to bring most IP activities under internal control, thereby creating the needs to form a proper IP department. This can be in the form of a single part-time employee in small firms to a proper department or division with dozens of employees. In the largest corporations, there will typically be a vice president dedicated to the management of the intellectual capital of the firm.

DRAFTING STYLE: DELIBERATELY GOOD OR DELIBERATELY DEFICIENT?

According to Article 78 EPC (requirements of the European patent application), a patent application filed at the EPO must comply with the following criteria:

- enclose an official request for grant;
- provide a description of the invention;
- have at least one claim;
- display any drawings referred to in the description of the claims; and
- provide an abstract

In Europe, as opposed to the US, there is no explicit obligation on the part of the applicant to cite the relevant prior art in the description (patent and non-patent literature) though Rule 27(1)(b) of the EPC alludes to the usefulness of doing so: 'indicate the background art which, as far as known to the applicant, can be regarded as useful for understanding the invention, for drawing up the European search report and for the examination, and preferably, cite the documents reflecting such art'.

Though formally complying with these criteria, the drafting style itself may vary quite considerably according to the experience of the applicant. A newcomer or improperly trained applicant is likely to produce a 'bad' application. A well drafted application, fulfilling all major requirements of the patent law, will ensure a smooth progress throughout the procedure, when drafted to the letter and exactly according to the requirements of the EPC. A maximum of one office examination communication will be issued, resulting in a reply from an accommodating attorney acting for the applicant and that remedies any defects and eventually leads to a communication from the EPO of intention to grant. In the best of cases, the applicant will be issued only one communication announcing a direct intention to grant, or 'direct grant'.

Imperfectly drafted patents might be the result of either a deliberate behaviour, or of a lack of skills (clearly insufficient knowledge of the requirements). The latter typically occurs with novices, individual inventors, irregular applicants, and SMEs acting without the use of an experienced representative.

There are cases where the applicant deliberately drafts the application so that it does not meet the requirements of the patent convention. The consequence of these practices is to increase the number of communications between the EPO and the applicant, and hence the length of the grant procedure. One such example is a communication from the attorney relative to the application EP03388083.2. Here is a quote from a communication from the applicant's representative: '... the present European patent application ... is deliberately drafted without fulfilment of all provisions of the implementing regulations ...'

The drafting style may also be deliberately complex in order to create a smokescreen that aims to hide the actual—or actual lack of—invention. However, a complex drafting style may sometimes be unavoidable due to the nature of the innovation and the technical field concerned.

The official working languages of the EPO and hence in which an application may be filed are English, French, and German. However, given that these are not the official languages of all EPO member states, nor of all countries from which applications originate, the drafting style will further depend on the language skills of the person(s) preparing or translating the application. Applicants drafting in their own language have many advantages over those who are not and will therefore be able to draft the application correctly with greater ease. However, no translation costs are borne until the final grant and validation of the patent in the elected designated states.

Description Part of the Application

A 'good' patent is a patent where the description is precise and to the point: it is easy to deal with and will probably be processed rapidly by the patent office.

The EPC actually requires 'clarity' as a necessary condition for the granting of a patent. The patents that are more complicated to understand, and hence to examine, are the patents that look like an operating or instruction manual, a research paper, or a PhD thesis, or that are highly complex, with a multitude of examples, drawings, and permutations.

Numerous applications are drafted in a style reminiscent of a technical operating manual, which makes the understanding of the alleged invention difficult to comprehend. One example is provided by the filing US5146591 at the USPTO: 'Dynamic information management system utilizing entity-relationship information model in which the attribute is independent of an entity'.

For the patents that look like a research paper it would appear that the body of the application has been prepared and written by the researcher, while the claims and other formal aspects have been added by the patent attorney. See, for instance, patent EP933428: 'Method for Increasing the Content of Trehalose in Organisms Through the Transformation Thereof with the DNA of the Trehalose-6-Phosphate Synthetase/Phosphatase of Selaginella leidophylla'.

Few applications and/or grants are provided with brief or little details. Such applications pose a high degree of risk for the applicant, or provide a weak scope of protection, insofar as in function of the available prior art, there are few possibilities for amendments to overcome objections of lacking novelty and/or inventive step. Moreover, such patents will probably have a limited scope of protection if granted and will therefore be easy to circumvent for competitors. The shortest EPO patent granted is probably EP0315714 which at filing comprised of 129 words, 664 characters, and one claim on a single page: 'Using of engine house of airplane with running surface for creation of effect of Magnus for VTOL'. An even shorter patent application is EP123058: 'Die Bogenbeute', where the description and the claims comprise a total of 57 words and 292 characters.

Number of Claims

The requirement for patentability is that a patent application must have at least one claim (Article 78 EPC). Failure to comply with this provides grounds for refusal by the formalities section of the EPO. Most patent applications are composed of several claims. A rather low percentage of applications have only one claim. At the EPO this fraction varies between 0.6 percent and 1.0 percent of the total number of applications filed. Examples of granted patents and patent applications with only one claim are: EP1262415: 'Briefumschlag mit Aufreiss-System'; EP0971722: 'An Aphrodisiac for Application to the Sexual Organs'; EP1061048: '*Distributeur automatique de l'eau potable ménagère et de l'eau pétillante avec unités de traitement de l'eau*'; EP1167119: 'Support for Clothes in Caravans'; US20030152907: 'Process of Love'; and US20040005535: 'Process of Reincarnation'.

Claims may be either independent or dependent. An independent claim is a claim that stands alone and defines a certain scope of protection to be sought. A dependent claim is a claim that refers to an independent claim and which further delimits the scope of the independent claim by additional characterising features. Thus, the dependent claim should be read as comprising *all* the features of the independent claim *plus* those recited in the dependent claim itself. A dependent claim will therefore inevitably have a narrower scope than that of the independent claim to which it refers. For example, the independent claim 1 of the patent application EP1544324 reads: 'A cleaning solution, comprising a mixture of a first compound, a second compound and water', which can include a solution with alcohol, water, and soap, whereas the dependent claim 2 further defines 'the cleaning solution of claim 1, wherein the first compound is selected from the following group consisting of oxalic acid, malonic acid, tartaric acid, propene-1,2,3-tricarboxylic acid, sulfosuccinic acid, oxalacetic acid, methylenesuccinic acid, succinic acid, citramalic acid and malic acid'.

There are four categories of claims: process, product, use, and apparatus for manufacturing the product. Applications with a few independent claims and a limited number of dependent claims would constitute a 'good' patent. The EPC prescribes excess fees for applications having more than ten claims at filing. Nevertheless, as illustrated in Chapter 7, the average number of claims per patents has nowadays risen to a number closer to 20 (see Chapter 7).

The overall number of claims, including independent claims and dependent claims, often witnesses a deliberate strategy by the patent application drafter, despite the fee-based incentives in place in order to limit this number. In the EPC, excess claims fees are payable for the eleventh and further claim at filing. At the USPTO, there are excess claims fees for more than three independent claims, for more than 20 total claims and for multiple dependent claims. For applicants willing to pay excess claims fees, the overall number depends on what their filing strategy is, on the complexity of the invention, on what fall-back positions they want to include and on many other factors. The number of claims is, in addition, also closely linked to the breadth of claims, e.g. a very broad independent claim may become considerably more limited in scope if combined with one or more of its dependent claims.

In fact, there are indicators that the independent claims are becoming broader, whether deliberately or not, which can be shown by an increasing percentage of applications not meeting the requirements of novelty and/or inventive step at filing. This percentage can, for example, be measured by the percentage of applications with a search report which has at least one citation of the 'X' or 'Y' category. This citation category indicates that at least one independent claim is anticipated by (or lacks inventive step over) the document to which the X/Y citation refers. A citation of the 'Y' category should also be considered in combination with any other 'Y' citations concerning the same claim.

In the process of preparing the search report, examiners will carry out a search for the prior art that is most relevant for making an opinion on the patentability of the claims on file. They will therefore not make a documentary search for all prior art relating to the subject disclosed in the patent application. The relevant documents that the examiners retrieve will then be cited against one or more claims of the application under scrutiny, and allocated categories according to the degree of relevance of the disclosure in the prior art document to novelty 'X', inventive step 'X' or 'Y', if the teachings of two or more documents are to be combined, or 'A' if the disclosure is only for relevant but non-damaging prior art. The category 'D' applies to documents cited by the applicant in the description. Other citation categories are also possible.

Currently, over 80 percent of PCT and about 75 percent of EP search reports issued by the EPO for applications filed in 2004 comprise 'X' and/or 'Y' citations, indicating that less than 25 percent of all PCT or EP applications at the time of filing meet the requirements for novelty and/or inventive step (Hammer 2005). This increase in X and Y citations is probably the consequence of the new patent strategies that consist in inventing around, creating patent thickets etc (see Chapter 4), which finally end up in a snowball effect: more and more applications are applied with a clear inflation in the number of claims.

Moreover, depending on the kind of innovation for which protection is sought, there may also be independent claims in different categories, such as process or method, product, use, apparatus etc. The inclusion of independent claims in more than one category will, in itself, drive up the total number of claims to more than just the sum of the number of independent claims, insofar as each independent claim will have a number of sub-claims comprising more limiting features dependent thereon.

The FESTO decision (see Box 6.1) in the USA most definitely had an important impact on the average number of claims, both independent and

---

Box 6.1. The FESTO case (2002)

The FESTO decision by the US Supreme Court essentially concerned the interpretation of the so-called 'doctrine of equivalents'. This relates to how literal the wording of a claim should be understood, and the determination of how broad a scope a claim may have. FESTO had as an effect that patents as a result of the examination procedure would be granted with a narrower scope than desired by the applicant. This left open a gap wherein competitors to the patentee could circumvent the patent and avoid possible infringement by simply making insubstantial changes to his or her product (i.e., by applying equivalents that were not covered by the patent concerned). In order to prevent this circumvention, patent applicants therefore began a practice of including claims to all possible foreseeable equivalents in their applications, thereby initiating a cycle of rapid inflation in the average number of claims in new applications being filed.

dependent, in applications subsequently filed (see for example Israelsen et al. 2002 and Miller 2002). There is an increasing number of patent applications with many independent claims associated with a large number of dependent claims. A knock-on effect is still being observed in patent applications filed worldwide as there is an increasing trend that applicants filing in more than one patent jurisdiction prefer to prepare one and the same application that can be used for filings at the various selected patent offices.

At the EPO a new Rule 29(2) of the EPC was introduced in 2002 in order to limit the number of independent claims in the same category. Some applicants nevertheless seek ways of circumventing it in order to maximize the potential scope of protection. Except where absolutely necessary, a European patent application may not contain more than one independent claim in the same category (product, process, apparatus, or use: Rule 29(2) of the EPC).

Patents may also be filed with a large number of independent claims and complex inter-relation of dependent claims. These patents are not likely to comply with the requirements for unity of invention (Article 82 EPC: 'The European patent application shall relate to one invention only or to a group of inventions so linked as to form a single inventive concept'). If an application comprises claims of different categories (e.g., process and apparatus) because of the claim breadth and the novelty/inventiveness of the common features linking the independent claims, the examiner may declare a lack of unity of invention. Obviously, the more independent claims made and the more disparate the features included are, the higher the likelihood of this being the case (see Box 6.2).

Some patents fall under the category of 'MEGA' applications. They comprise an extremely high number of claims, several hundreds, if not thousands. These patents have alternatively been labelled 'Jumbo' applications (Michel forthcoming), 'nasties' or 'supernasties' (Milne 1991). Particularly extreme is a number of series of applications filed by Shell (the publication WO03040513 is composed of 8958 claims); Enventure; Hawaii Biotech (the publications WO2004011423 and US20050004235 are composed of 1015 and 2529 claims, respectively); and Angiotech, which applied for several MEGA patents (WO2005051444 with 19,368 claims; WO2005049105 with 17,517 claims, and WO2005041451 with 13,305 claims).

It is a deliberate drafting practice by some applicants to file such applications with the reasoning being that it is up to the examiner to clean up the claims, hence posing an unwanted burden on the patent office. For both the search and the substantive examination, such applications can present virtual 'needles in a haystack', as significant extra efforts are required in order to ascertain where the invention, if any at all, resides. These applications create a disproportionate degree of uncertainty to competitors and the public at large, since it is very difficult, or nearly impossible, to predict what scope of protection will finally be granted, if any at all.

Box 6.2. The RIM case, part 2: NTP's claim drafting style

The *RIM vs. NTP* case is described in Box 4.3. It suggests that the US patent system might have wrongly (according to the USPTO) granted several patents to NTP, inducing a settlement of US$612.5 million in favour of NTP and at the expense of RIM. Some newspapers have further underlined the fact that the USPTO might not have enough resources to tackle the overload of patent applications properly.

The granting decision by an examiner does not only depend on the technological invention itself, but also on the way the claims and description are written. There are indeed specific patent claims strategies that might be adopted in order to increase the chance of being granted. Some of these strategies are defined by Edward J. Brooks and Charles R. Ware (2006).

NTP is taken by the authors as an example that reinforces the importance of making wise 'claim choices'. Patent prosecutors must indeed consider how to allocate claims in each patent applications, and what type of claims to use—and how many of each to use. In the US patent system, the claim typology includes:

- apparatus claims (with structural limitations, or with functional limitations);
- method claims (of fabrication, assembling, or operation);
- system/network claims;
- Beauregard claims (for an inventive computer program);
- Lowry claims (for inventive data structure); and
- means claims (one or several elements expressed in mean-plus-function language).

The reliance or not on each of these claims depends on the strategy adopted by claim drafters.

In 1995, three inventors obtained the US patent no. 5,436,960, which claimed systems and methods for transmitting email by radio frequencies to portable receiving units. The '960' patent was composed of 89 claims, which included four independent system claims and four independent method claims. None of the claims were Beauregard claims or means claims. The patent contained 24 pages of text, including nine pages of software codes. As the patent was too broad, four 'continuations in parts' applications had to be filed (at the EPO it would be called a 'divisional application'), yielding more than 1500 additional claims, all assigned to NTP.

In 2001, NTP sued RIM for infringement. At trial, RIM asserted that part of its accused network was not within the US and therefore the entire network was outside reach of the patent infringement statute (the BlackBerry network relay, one of the allegedly infringing element) was in Canada. The court eventually rejected RIM's defense and the jury found that RIM infringed upon 14 of the patent claims asserted by NTP, awarded US$54 million to NTP and issued a permanent injunction against RIM. RIM appealed.

It is interesting to observe that the trial focused on geographical issues, instead on the validity of the claims, which should probably not have been granted at the first place (see Box 4.3). Brooks and Ware praise the drafter strategy:

although the patent discloses software code and contains method claims, it does not look like the drafter tried to claim software; rather it looks like the drafter used the code to enable method claims. It seems that the drafter attempted to use system claims to cover the structure of a wireless e-mail network and method claims to cover the function of this network. The drafter further intended to gain broad coverage by using numerous independent claims to claim many different independent embodiments.

Of the 14 infringed claims, some were system claims and some were method claims. (See Box 7.2 and Box 7.3 for an illustration of the consequences on RIM's behaviour regarding its patenting strategy.)

(Brooks and Ware 2006)

This is stated very succinctly in the description part of GB1388517: For these reasons, patent work is likely always to be a complex and somewhat 'hit-or-miss', process and the job of the Patent Examiner, as an everlasting 'Searcher for new needles in a technical haystack', growing larger every year, must be about the most 'soul-destroying professional occupation in Science or technology', and likely to reduce anyone to a nervous breakdown in two years, since the meaning attributed to just a few words in a main claim, can change its whole validity in relation to prior art.

Scope of Claims

Beside the number of claims, their size may also vary substantially. The shortest and narrowest known claim in a granted patent is probably that of claim 1 in the patent US3156523 issued by the USPTO on 10 November 1964 to Glenn T. Seaborg and assigned to the United States Atomic Energy Commission. The claim reads: 'Element 95'.

Applications with broad claims will typically try to cover as much 'territory' as possible in the technological space. This can be also expanded by including a large number of independent claims. Further adding a number of dependent claims covering multiple alternatives provides fall-back positions.

For patents with a broad scope, and/or with many complex alternatives, risks arise with respect to subsequent challenges in opposition or litigation proceedings if these claims are granted in unamended form, as damaging prior art may pass undetected by the patent office but retrieved later by opponents (see sections 6.3 and 6.4).

A detrimental effect of applications comprising unduly broad and/or numerous claims, and in addition often very lengthy descriptions, is that when published they pollute the prior art and increase the overall uncertainty by making patent searching more difficult and rendering the evaluation of patentability more complex (Philipp 2006). Moreover, during the examination process of the application, there is an issue of provisional patent protection, the actual scope of which is uncertain until the patent is granted (see Figure 6.2). This can therefore have a stifling effect on further research and bias market competition in an unwarranted manner.

In particular in the field of organic chemistry, the filing of so-called Markush claims is a common practice, named after Eugene A. Markush who was granted the patent US1506316 in 1924 for 'Pyrazolone dye and process of making the same'. In one particular case, a patent application was filed with a claim to a chemical structure that was calculated to encompass 10^{60} possible compounds (see Milne 1991; Franzozi 2003).

A Markush claim refers to a claim to a chemical structure with multiple 'functionally equivalent' chemical entities in one or more parts of the compound. The more parts of the compound with alternatives and the more alternatives per part, the higher the total number of the resulting different

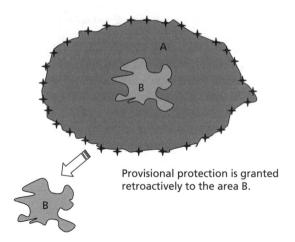

Provisional protection is granted retroactively to the area B.

Figure 6.2. Scope of patent applications at filing and at grant

Notes: A = scope of claim as filed and published; B = scope of claims as granted.

possible chemical compounds. For example, a Markush structure for a compound having a core with ten variable parts, each part provided with ten alternatives will correspond to a total of 100 different compounds.

Number of Pages

The EPC contains a number of regulations concerning the formal layout of the application documents (e.g., in terms of the page layout and the like). These are found in the Implementing Regulations of the EPC. One requirement, however, that is not indicated therein is the font type, which may have an important impact on the final number of pages (see for instance: Rule 32 EPC: Form of the drawings; Rule 33 EPC: Form and content of the abstract; Rule 35 EPC: General provisions governing the presentation of the application documents).

A single page is an absolute minimum. In fact, such applications will be published with a minimum of two pages: a front page with bibliographic information and the abstract, and a second page with the description and the claim(s). Any search report will, of course, take up some more pages upon publication.

The fee structure of the EPO foresees an excess fee for applications comprising more than 30 pages at the time of grant (As of May 2006). Hence, a large proportion of applications will contain 30 pages or less. If the number of pages in a patent is greater than 30, supplementary fees are requested by the EPO, but only at the time of grant and printing. Therefore, an initially larger application, possibly with a lack of unity, that has induced the filing of divisional application(s), may have had a certain number of claims cancelled and description pages excised, and may result in an application within the threshold of 30 pages.

Article 82 of the EPC stipulates that the application shall relate to one invention only or to a group of inventions linked to form a single general inventive concept. Depending on the drafting of the claims, an applicant may therefore intentionally or accidentally induce an objection of lack of unity from the examiner on the grounds that the claims a priori (i.e., before a search is carried out) or a posteriori (i.e., after a search has revealed relevant prior art) present a number of different alleged inventions. Upon the payment of further fees, a search for the other alleged invention(s) will be carried out. In the European examination, however, the applicants will be invited to decide which of the alleged invention(s) they wish to have examined as the main file. Should they wish to have the other alleged invention(s) examined as well, then they must file one or more divisional application(s), each corresponding to one alleged invention. This is subject to the conditions laid out in Article 76 EPC.

There are some applications with hundreds or thousands of pages, not just in the claims section but also devoted to the description and the drawings and/or biological sequence listings. Such applications impose severe difficulties in detecting possible added subject matter in amended claims—where do the amendments find support in the description part? Moreover, large applications impose significant difficulties in their physical and intellectual handling. Last but not least, such large applications will form part of the prior art when published, leading to difficulties in the electronic handling, in searching and so forth.

Complexity of Invention

Dack and Cohen (2001) suggest that without a minimum level of complexity there is generally no invention. 'Simple' (or obvious) patents will often be disposed of at an early stage in the procedure, as prior art is easy to find and communicate to the applicant, who then normally takes the obvious conclusion and withdraws.

Applications with a degree of complexity that is average to normal, with a clear statement of a problem and its solution, will often belong to the 'good patent' category. Thus, the requirements of EPC Rule 27 would normally be fulfilled; the description shall:

- specify the technical field to which the invention relates;
- indicate the background art which, as far as known to the applicant, can be regarded as useful for understanding the invention, for drawing up the European search report and for the examination, and, preferably, cite the documents reflecting such art;
- disclose the invention, as claimed, in such terms that the technical problem (even if not expressly stated as such) and its solution can be understood, and

state any advantageous effects of the invention with reference to the background art;

- briefly describe the figures in the drawings, if any;

- describe in detail at least one way of carrying out the invention claimed using examples where appropriate and referring to the drawings, if any; and

- indicate explicitly, when it is not obvious from the description or nature of the invention, the way in which the invention is effectively subject of industrial exploitation.

6.3. A Step by Step Interaction with the EPO

The EPO is charged with the tasks of implementing the procedures laid down in the European Patent Convention (EPC). For this purpose, the EPO is organized in a Receiving Section, Search Divisions, Examining Divisions, Opposition Divisions, a Legal Division, Boards of Appeal, and an Enlarged Board of Appeal.

In addition to the EPC, to support the work of the Formalities Officers and Examiners charged with the tasks related to filing, search, examination, and opposition there are a set of 'Guidelines for Examination in the European Patent Office'. These provide guidance in the interpretation of the relevant Articles and Rules of the EPC and examples of how to deal with specific situations that may arise during the procedure. They are based on the experience of the examiners and of the Board of Appeal.

Additional tools, for the examiners in particular, are a set of Internal Instructions with more details of how to proceed with search and examination, and a regularly updated reference book on the case law emanating from the Boards of Appeal.

Once an application is filed at the EPO it starts the procedural stages illustrated in Figure 6.3. If the content and format of the patent comply with Article 78(1) EPC it is subject to a search for relevant prior art by the patent office. The search aims at finding the most relevant state of the art. It corresponds to all the information, relevant to the patentability of the invention, made available to the public (in a written or oral form) prior to the date of the priority filing (Article 54(2)).

The search is performed by the examiner and is mainly performed through the EPO database EPODOC, which contains over 55 million prior art

Figure 6.3. The European patenting procedure

documents. Other databases, specific to the technological domain, and the Internet may be consulted as well. The information contained in the EPODOC database is classified within approximately 70,000 defined classes (see the International Patent Classification (IPC) available on the WIPO website). A further refinement of the IPC is the European Classification (ECLA), which is developed and maintained by EPO examiners, and which contains more than 220,000 defined classes. The examiner assigns one or several 7+ character classes to the application.

Since 2004 for PCT applications, and since July 2005 for direct applications at the EPO, the search report is accompanied by a non-binding opinion regarding the prospect of the application to be granted, a kind of pre-exam report. The search report and the opinion become public at the date of publication of the application (30 months after the priority filing for PCT applications and for national and EPO direct application it takes18 months after the priority filing according to Article 93 of the EPC).

NOTICE OF SEARCH REPORT AND FIRST WRITTEN OPINION

Once the search report and the first written opinion on patentability are sent to the applicant, she or he has to make a decision on whether:

- to proceed further and request a substantive examination—optionally with amendments to the application before commencement of the examination procedure; or
- to withdraw the application either with or without notice to the EPO.

A decision to proceed further depends on factors such as the contents of the search report and written opinion (i.e., the likelihood of arriving at an eventual grant and its potential scope), the viability of the innovation, possibilities for marketing, expected profits, the underlying patenting strategy (see Chapter 4), and the tenacity of the applicant.

If the applicant decides to proceed further at the EPO, a formal request for examination must be made and the required fee paid within the prescribed time limits (Article 94(2) of the EPC stipulates that a request for examination may be filed within a period of up to six months after publication of the search report). For applications filed within the new Extended European Search Report (EESR) procedure relating to applications filed after 1 July 2005, use may be made of a waiver, so that the file automatically enters the examination phase upon issuance of the search report (EESR: 'As from 01 July 2005, all European applications are subject to a search for relevant prior art and the search report will be issued together with a written opinion on patentability' (EPO 2005a: 435–40, no. 7)).

With the request for examination, the applicant may on his or her own will also file an amended set of claims, or other amendments to the application, provided they do not contravene the requirements of the EPC concerning added subject matter.

An active decision to withdraw the application can be communicated to the EPO, after which any outstanding rights are abandoned and official notice will be made that the application is dead. However, it may be of more use if the applicant intentionally leaves the competition in doubt concerning the final outcome and let it lapse. Such a situation generally happens when the applicant does not pay the 'examination fees'. This will after a certain time be officially communicated by the EPO, and the application will be registered as 'deemed withdrawn'.

PROSECUTION OF SUBSTANTIVE EXAMINATION

Once the request is made for a substantive examination, the examiner analyses the application and is in charge of written communications to the applicant, as well of the analysis of the applicant's replies and amendments. The examiner ultimately recommends grant or refusal of the application. The examiner being in charge is called the 'first' examiner, a second examiner is charged with assuring the correctness of the formal requirements of the recommendation and the form of the final text of the granted patent. The examining division is headed by a chairman who presides over the decision to grant or refuse, and in the event that there is disagreement between the members of the examining division carries out a detailed check of the application and decides whether to allow the refusal or grant of the patent, or to send it back to the first examiner with recommendation for further action.

The examiner devotes a special attention to the claims. They define the scope or extent of the protection of the patent and are the most important part of the application. The examiner also takes into account the drawings and the description, as they are used to interpret the claims.

In order to be granted a patent must fulfil the following criteria:

- there must be an invention;
- the invention must be new (Novelty, Article 54 of the EPC);
- the invention must involve an inventive step (Article 56);
- the invention must be susceptible of industrial application (Article 57);
- the EPC does not define an invention but Article 52(2) of the EPC contains a non-exhaustive list of things which shall not be considered as inventions: discoveries, scientific theories, mathematical methods,

aesthetic creations, schemes, rules, and methods for performing mental acts, playing games or doing business, computer programs, and presentations of information.

Further important requirements for patentability are that the claims must be clear and concise and supported by the description (i.e., that a person skilled in the art must be able to understand the exact scope of the claim as it is written and that the wording of the claim is reflected in the description part of the application). Moreover, the application as a whole must be drafted in such detail and in a sufficiently clear manner so as to allow a person skilled in the art to carry out the invention.

In the few cases that the relevant requirements are met at the time of entering examination, a direct grant is possible and the examiner issues an 'intention to grant' as first office action. A speedy conclusion of the procedure may be reached, only depending on the applicant's approval of the grant, filing of translations of the claims in the two other official EPO languages, and payment of the relevant fees (see below).

Otherwise, a cycle of office actions and applicant replies is then commenced. The overall time between filing and completion of the procedure is not given, as both the applicant and the patent office have an influence on the speed of the procedure (see below).

Internal efficiency at the EPO plays an important part, such as the organization of the workflow between and within departments but most important is the workload and staff capacity situation which is a determinant factor in the size of any backlogs.

Generally speaking, the longer the overall procedure, the longer the maintenance of the uncertainty linked to the application. This may have a profound influence on important decisions to be taken by the applicant and by competitors. It may therefore be useful for an applicant to have a prolonged procedure, as potential competitors may be dissuaded from entering the area covered by the scope of the claimed but not yet finally approved invention. This is often the case in industries with a long product development time, such as the pharmaceutical industry. On the other hand, in industries with short product life cycles, a speedy conclusion is more desirable (e.g., for selling the patent rights, assuring licensing fees, and the like).

A fast sequence of events is less and less common. Figure 6.4. shows that about 20% of all applications filed in 1999 were still pending in early 2006. Insofar as a great majority of applications do not meet many of the requirements of the EPC upon filing or upon entry into the examination stage, and these deficiencies are not all corrected with one office action and one applicant reply only. The normal procedure may be deviated from by particular actions on the part of the applicant but also by the patent office.

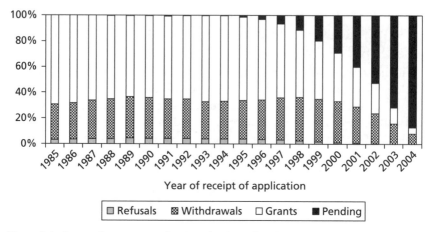

Figure 6.4. Status of past EPO applications, by date of applications, as of January 2006
Source: EPO (http://ep.espacenet.com).

FAST TRACK

The applicant has a number of means available that may help in assuring a quicker procedure, including the following:

- File the application under the PACE programme: the EPO ensures that the search report for an EP first filing will be issued within six months after filing (RASE, accelerated search report), and as soon as possible for all other EP search filings. In case of a PACE request for accelerated examination (RAEX) the EPO will make every effort possible in order to issue the first communication within three months after receipt of the request. For further communications, these will be issued within three months of receipt of the applicant's reply. No justification is needed, there is no additional fee, and the request is not publicized.

- File an early request for examination (Article 96(1) EPC).

- File amendments on the applicant's own volition upon issue of search report and written opinion, before entering the examination procedure.

- Reply to any office action within the statutory time limits.

- Assure payment of any required fees at or before the deadline.

- Request and obtain informal (telephone) interview(s) with the examiner to solve minor issues.

Depending on the office workload and that of the individual examiner, the procedure might by itself be quicker than average, for example if the examiner can deal with replies from the applicant immediately after receipt.

Other means that can be used in order to come to a speedy conclusion is the option of oral proceedings, which may be at the request of the applicant or at the initiative of the examining division. In any event, the oral proceedings can only be called by the issue of an official summons by the examining division. All three members of the examining division must agree to the contents of the summons as well as to the written decision issued after the proceedings, if held. Such procedure allows to reach a final settlement between the examiner and the applicant, instead of pursuing with a large number of communications. A 'threat' of oral proceedings is often a standard clause in communications from the both the applicant and the examiner.

SLOWER TRACK

The applicant, on the other hand, has a number of means available that may slow down and draw-out the procedure. These include but are not limited to the following:

- Choose the PCT as filing route. The earliest date for entry into the European Regional phase is 30 months after the priority filing.
- Request extension of the time limit to respond to any and/or all office action communications. Typically, the statutory four months' limit can be extended by at least two months. Further extensions can be obtained under certain conditions.
- Request that no office action is taken. In this case a valid reason must be given, such as serious illness, death, or bankruptcy.
- File sequential divisional applications. This is provided that the parent application is still pending. Keep divisional applications pending as long as possible.
- If possible, appeal any decision issued by the EPO.

A divisional application is subject to Article 76 EPC and is intended for seeking patent protection for subject matter which normally would contravene the requirements for unity of invention of the parent application. The extent of protection sought must not exceed that of the parent application as filed. Its date of filing will be the same as that of the earlier application. The filing of divisional applications is sometimes used to deliberately delay the conclusion of the procedure.

In some areas of technology, such as in the pharmaceutical field, applicants are not often at all interested in a quick procedure as they, at the time of filing, usually do not know which of the chemical entities claimed has the highest activity, the least toxicity, can be formulated into a drug, and is therefore subject to the approval of the authorities. They therefore frequently need a significantly longer processing time. As the bulk of the cost occurs once the patent is granted, applicants have no incentives to reach a fast granting decision (see section 7.2).

Though not always desirable from the applicant's point of view, the office and examiner workload can also contribute to a delayed procedure. Bearing in mind that delays are not evenly spread throughout all technical areas, a deliberate drafting of the application so that it is likely to be dealt within an over-loaded technical area might also help in assuring a longer procedure.

THIRD PARTY OBSERVATIONS

The examination procedure is essentially a negotiation between the applicant and the office, but there is a limited possibility for third parties to intervene during the procedure. By virtue of the electronic file inspection, any interested party has the possibility of monitoring the progress of individual applications and if direct intervention is deemed useful, there are provisions in the EPC thereof (Article 115 EPC: 'Observations by third parties').

These observations take the form of a written part which may give grounds for why a patent should not be granted and may also include prior art in support thereof. This prior art may be the same as already known but typically includes further documents that have not been cited by the applicant or retrieved by the search examiner in the search report. Though the patent office is not under obligation to include these observations in its considerations, their relevance to the case will in actual practice always be considered.

GRANT

The decision of the granting of a patent by an examiner is sent to the applicant as an intention to grant (IGRA), which can in turn be accepted or refused by the applicant. If a quick publication of the grant is sought, this can be assured by replying to the IGRA as soon as possible with a fast payment of fees. A slower process is also possible, as witnessed by the following options available to the applicant:

- Reply to Rule 51(4) communication as late as possible (communication that the EPO intends to issue the grant with an invitation to pay the fees for grant and printing, and to file translation of the claims into the other two official languages).

- Request amendments or correction of errors according to Rule 51(5) within the time period set out in Rule 51(4). Deliberately assure that the examining division will not consent to these, so that a communication under Rule 51(6) will be issued (the applicant has the possibility to submit observations or amendments in the event that the examining division has not consented to a former amendment or correction).

- Subsequently use the normal full period under Rule 51(6) for submitting amendments and for filing the translation of the claims and for paying the grant and printing fees. This period is non-extendable.
- If any renewal fees are due and required to be paid under Rule 51(9), make use of the full six months grace period for payment.
- If no renewal fees are due, after replying to the Rule 51(6) communication of the grant is not likely to be issued earlier than 5 months from the date of the Rule 51(6) communication under Article 97(5). This is a result of the time limit for the filing of a translation of the claims into the two other official languages, and to allow for internal processing at the EPO.

Applying these steps, publication of the grant is likely to occur after one year. Once the patent is granted by the EPO it must be validated in each desired National Patent Office. This induces the payment of translation, validation fees and renewal fees for the desired length of protection. In case of a refusal by the examiner, the applicant can make an appeal before the EPO.

OPPOSITION

An opposition to a grant may be filed by third parties before the EPO. In most cases, this procedure, which is carried out by an opposition division of the EPO; terminates with oral proceedings; and is open to the public. It is a centralized EPO procedure that allows the public to challenge the examining division's assessment of patentability. The opposition division is composed of three members, at least two of whom did not take part in the examination proceedings (it usually includes the first examiner). According to EPC Article 99, an opposition must be filed within nine months of the mention of grant being published. It can be filed by any person or institution. The opposition process allows for the introduction of disclosures which may not have been available to the examining division. These disclosures can be either oral, related to prior use, or printed on proprietary disclosures, such as technical manuals.

The outcome of an opposition procedure before the EPO will lead to the revocation, amendment, or maintenance of the European patent in all designated states. It is on average 1/3:1/3:1/3, meaning: one-third of patents are maintained in un-amended form; one-third are maintained in amended form; and one-third are revoked. In the event of a maintained or amended patent, a new specification will be published.

Opposition proceedings are rather costly due to attorney fees, expert witnesses, retrieval of vital evidence for invalidating a patent and so forth. However, oppositions are much cheaper than litigations that may proceed before any national court, after the patent is granted and put in force in various EPC Member states. In 2002, 2200 oppositions were filed before the EPO, and

3100 in 2004. Although the absolute number of oppositions is increasing, the rate of oppositions in relation to the number of published grants is decreasing and currently corresponds to about 5 percent of all patents granted.

Opposition against a granted European patent may be filed on the basis of the following grounds:

- lack of novelty;
- lack of inventive step;
- non-patentable subject matter;
- lack of industrial applicability;
- lack of or insufficient disclosure;
- inadmissible extension of the patent disclosure beyond the content of the application as originally filed.

Box 6.3. Henkel's opposition strategy

Oppositions reflect a clear strategic behaviour of firms. Opponents are typically rivals of the patent holder who seek to have the patent revoked or narrowed. Historically, about 5 to 8 percent of all patents granted each year by the EPO have been challenged. The frequency of oppositions and their success rate can be taken as evidence supporting the hypothesis that an opponent has a strong capability of getting rivals' patents revoked. In combination with low success rates, frequent oppositions would tend to point to strategic behaviour or low cost of opposition. In this respect, Harhoff (2006) provides ample evidence that firms differ remarkably in how successfully they handle opposition challenges. The author shows that in cosmetics, Henkel attacks almost twice as many patents as it owns. No other firm uses opposition so frequently in this field.

Henkel attacks more frequently than other firms do (e.g., l'Oreal, Unilever, and P&G), and there is no notable reduction in the company's opposition success that would point to a trade-off between frequency and success of opposition. The simplest economic explanation for using opposition frequently is to employ it for strategic purposes, such as to deter rivals. But that should be accompanied by a reduction in the opponent's success in having rivals' patents revoked. As this is not the case with Henkel, one has to conclude that some firms are simply better at opposition than others. Deeply ingrained organizational capabilities in patent documentation and IP management may be a decisive determinant of opposition success. Indeed, Henkel has had a long history of supporting investments in patent documentation and IP management.

One of these investments took a particular form—Henkel's participation in the *Internationale Dokumentationsgesellschaft für Chemie mbH* (International Documentation Corporation for Chemistry (IDC), created in 1967 and closed in 1997 due to the rise of Internet and competing service providers). Members of the IDC derived competitive advantage from superior knowledge concerning the state of the art. This had direct implications both for the drafting of patents (which were less likely to be attacked) and for attacking rivals' patents in opposition. More comprehensive knowledge concerning the state of the art also enabled member companies to file 'stronger', more sustainable patent applications. Henkel is still a member of the Patent Documentation Group (PDG), founded in 1957, and which currently counts about 38 member companies in eight countries.

The evidence produced for an opposition may comprise new prior art or evidence of prior use, including statements of expert witnesses but will also very often comprise a review of the procedure and arguments against the patentability of the filed invention.

The filing of an opposition forms an important part of the interaction between competitors in a given technical area, but the absolute rate of filing oppositions may not necessarily reflect the true intensity of the competitive environment. Many patents are never opposed due to a perceived lack of value, and many oppositions are never filed but are replaced with a privately negotiated settlements. A similar situation is also known to apply to litigations. The average duration of an opposition is about two years and an additional two years must be taken into account for appeal cases.

6.4. **Typology of Filing Strategies**

The previous sections have shown the various dimensions of a patent filing strategy at the EPO. It requires applicants to select a route for patenting, to adopt a particular drafting style, and to choose appropriate actions when interacting with the EPO. The overall filing strategy, if not the result of a clear lack of skills or experience, is driven by the broad market strategy adopted for an invention.

Table 6.3 provides a typology of filing strategies. The choice of a particular strategy will depend on the broad intellectual property objectives of the firm, which in turn depends on its technological and competitive environment. Four broad patent strategies can be identified:

- good will and fast track;
- good will and slow track;
- bad will and slow track; and
- clear abuse of the system.

A good will and fast track strategy characterizes the patent filings for which the applicant expects a fast grant and a fair scope of protection. The fast track might be induced by an urgent need for protection or by a short technology life cycle of the invention. In the former case urgency might be requested by potential investors or venture capitalists, or simply be a condition for technology negotiations. An indicator of the relatively small number of fast track procedures is that less than 0.9 percent of all patents issued in 2005 by the EPO had a filing date in 2004 and about 8 percent had a filing date in 2003.

The degree of urgency will first influence the patenting route. If a 'good will and fast track' is the preferred strategy, the applicant will either make a direct application at the EPO or file a national priority and make a subsequent second filing at the EPO. In addition, the application must be 'well' drafted (i.e., it must

Table 6.2. Typology of filing strategies at the EPO

Filing strategy	Filing practices and behaviours:		
	Patenting route	Drafting	Interaction with EPO
Good will and fast track	• EP-direct; or • national priority followed by fast EP-second filing.	• good drafting;* • normal complexity; • relevant claims;** • formal; • concise description; • clear problem and solution.	• accelerated search and examination; • substantive examination filed with amendment of original file; • fast replies to communications; • fast payment of fees; • request earlier grant.
Good will and slow track	• Euro-PCT filing.	• good drafting;* • normal complexity; • relevant claims;** • formal; • concise description; • clear problem and solution.	• normal interaction; • apply the longest delays possible for replies; • wait for communication for amendments; • late payment of fees.
Bad will and slow track	• Euro-PCT; or • PCT filing in the US.	• deliberately deficient;* • highly complex; • too many claims; • long claims; • unclear description; • lack of unity.	• delayed interaction; • wait for communication for amendments of claims; • request maximum extension for replies; • late payment of fees.
Deliberate abuse of the system	• US priority followed by PCT.	• deliberately deficient;* • long list of prior art; • extremely complex; • high number of claims; • many independent claims; • long claims; • cross references between claims • unclear, long description • invention hidden • lack of unity • successive Divisionals each with slow prosecution.	• delayed interaction; • wait for communication for amendments of claims; • request maximum extension for replies; • file Divisionals and possibly Divisionals of Divisionals; • late payment of fees; • file an appeal on some decisions.

Notes: *Applications are qualified as 'deliberately deficient' if they do not meet all the requirements of the EPC on purpose. They therefore induce communications from the examining division; **'Relevant claims' stand for applications with a few independent and dependent claims. With too many claims the targeted scope is too large and there is potentially a lack of unity, which means that divisional applications are likely to be filed.

meet all the requirement of the EPC), with a clear explanation of the problem and its solution, and a concise description that would allow the examiner to directly understand and evaluate the invention.

The interaction with the EPO also matters. From the start, a request for accelerated search and examination can be requested. Then the applicant will avoid waiting for the entire 'legal' periods of time before replying to a communication from the Office (i.e., four months) or before paying the fees. If the search report suggests that one or several claims might have to be amended, the applicant can submit an amended version of her file for the substantive examination, instead of waiting for a communication by the examiner, formally requesting one or several amendments that were foreseen in the non binding opinion associated with the search report.

A 'good will and slow track' strategy is probably more frequent than the previous one. Basically it reflects an attempt to lengthen the procedural delay of a filing, probably to avoid the large expenses that have to be sustained once the patent is granted (especially if a large geographical scope is targeted), or because the applicant needs more time to assess the value of the patent (i.e., when performing a business plan and market survey). In this case the applicant will file for a PCT filing, which allows to wait for more than 30 months before entering the EPO regional phase. The drafting style would basically be the same as for the fast track strategy. Nevertheless, the applicant might already draft the application so as to ensure one or two communications from the EPO during the examination process. For both the replies and payments of fees, the applicant will wait for the latest time limit.

A 'bad will and slow track' strategy seems to become more and more frequent. Such behaviour characterizes applicants who want the broadest scope of protection that is possible to reach. There is no urgency and the uncertainty generated by the numerous claims and the unclear description of the filing affect the competitors as well. The applicant will try to delay the procedure for the longest period of time possible, to maintain a level of uncertainty on the market and/or to delay the costs that he or she would have to bear at the grant date, if granted.

This is a typical situation where the applicant will go through a PCT application, may be with a priority filing at the USPTO (the fees are lower, there is a grace period that basically allows to submit already published scientific papers, and continuous applications are allowed, which means that there is no formal requirement for the drafting of the document and that substantial changes can be made). The drafting style is unclear, with a large number of dependent and independent claims interacting with each others. This complexity, associated with an unclear description, leads to several potential communications from the examiner, requesting several amendments, or divisional applications. The applicant can then wait for six months to reply, and may make some additional new mistakes in the amended file.

The 'clear abuse of the system' strategy, although currently being performed by a minority of applications, appears to be growing fast. It reflects a definite will to hide an invention, or a claim, in an ocean of useless or low quality claims. It then becomes easy to wait for potential users of the technology, and litigate them for infringement. It is a typical strategy adopted by the so-called 'patent trolls' or 'submarines applications' (see Chapter 4), which often lead to the filing of MEGA applications, or highly complex ones, that might be subject to the request of several divisional applications, i.e. the office requests that an initial filing is split into at least two patents. They all keep the same priority number and priority date but will be allocated separate and different application numbers.

It is difficult in practice to measure the relative importance of each type of strategy in the total number of patent applications. The main reason is that behavioural characteristics are not easy to assess. According to examiners and applicants, it seems reasonable to say that the second and third strategies are the most frequent: 'good will with slow track' and 'bad will with slow track'. The slow track is justified by the high expenses that occur just after the grant (see Chapter 7 for the cost of patenting in Europe). The 'bad will and slow track' may be imposed by the competitors' behaviour of by the perceived uncertainty regarding the future commercialized form of the patented invention.

Avoiding imitation requires to block whole areas of technology. As illustrated in Chapter 4, it is apparently easy nowadays to 'invent around' (see the 12 rules to 'invent around' put forward by Glazier (2000)), and there are 'antidotes' for such practices (see, the antidotes of inventing around, also put forward by Glazier (2000)), which basically translate into a snowball effect. In this context of patent race, and given the high cost associated with grants, the multitude of patent filings would definitely be subject to a slow track strategy.

An indirect measure of these strategies is represented by the status of EPO applications over time, as illustrated in Figure 6.4. A small percentage of the applications filed in 1996 and 1997, about 5 percent, are still pending. Most of these applications have been pending for over ten years. Out of the applications filed in 1999, more than 20 percent are still pending, seven years later. This is the results of either well managed procedural delays, or a very high complexity of the application, or both.

The relationship between the number of claims (an indicator of complexity if a large number of claims are included in the applications) and the procedural duration has been recently illustrated by a working document of the EPO (see Lazaridis and van Pottelsberghe 2006). On average, from a baseline of about 11 claims per application, two additional claims induce an additional communication from the examiner to the applicant. And an additional communication leads to a procedural duration of one additional year, whatever the final outcome of examination process (i.e., granted, withdrawn, or refused). In other words, the number of claims influences both the amount of work supported by examiners (e.g., the number of communications) and the average length of the granting process.

6.5. **Summary**

- By filing a single application in one of the three official languages (English, French, and German) and by a single patent grant procedure it is possible to obtain patent protection in some or all of the EPC contracting states. It is also possible to bypass the EPO for an effective protection in several European countries. The European Patent System becomes complex and costly just after the grant, as a patent must then be validated, put in force and renewed in each national patent system, with its own legislation and its own fees structure.

- Applying for a patent requires the choice of a route for patenting; the selection or not of a professional representation; the adoption of a particular drafting style; the choice of an appropriate behaviour when interacting with the EPO; and readiness to support the corresponding expenses.

- The granting process includes various stages, from the search for prior art, to the substantive examination and the final decision to grant the application. After the grant, third parties have nine months to file an opposition before the EPO. Each of these stages include at least one official communication from the EPO to the applicant, and the payment of fees. The applicant may request a fast track or, conversely, they may adopt a behaviour that would substantially delay the granting process.

- A typology of four broad filing strategies may be used to characterize the applicants' behaviour: a good will with fast track, a good will with slow track, a bad will with slow track, and a deliberate abuse of the system. The chosen strategy will affect the patenting route, the patent drafting style, and the interaction with the EPO.

- A fast sequence of events is less and less common, insofar as many applications do not meet many of the requirements of the EPC upon filing or upon entry into the examination stage. The practices that consist in delaying the granting process may be explained by the high costs that must be incurred once the patent is granted.

- The deliberate abuse of the system includes a drafting style of the application that is 'deliberately deficient' (i.e., with a large number of claims and a complex description) and may induce (1) an unwanted burden on the patent office; (2) a disproportionate degree of uncertainty to competitors and the public at large, since it is nearly impossible to predict what scope of protection will finally be granted, if any at all; and (3) an unclear published prior art, leading to difficulties in the electronic handling, searching, and identifying the relevant prior art for future applications.

- The slow track strategy is witnessed by status of past patent applications. For instance, 20 percent of the patents applied in 1999 were still pending in May 2006. The increasing delays in the granting process are partly explained by the number of claims included in an application. On average, two additional claims induce an additional communication from the EPO to the applicant, which in turn leads to one additional year in the procedural duration.

7 Hot 'Patent' Issues: Quantitative Evidence

Bruno van Pottelsberghe

This chapter analyses the economic implications of four contemporary issues related to the patent system. The first section intends to shed some light on the economic implications of the arrival of a new type of institutions in patent systems: universities and public laboratories. The second section is dedicated to an evaluation of the cost of patenting in the European, US, and Japanese patent systems. It illustrates how the current fragmentation of European markets for technology leads to very high relative costs, and how it may affect the demand for patents. Section three presents a comparative analysis of operations in the three main patent offices. It shows that the higher relative costs in Europe are partly compensated by a higher rigour of the granting process. The fourth section describes the current European policy changes, and their potential economic impact, that are put forward to improve the integration of the European market for technology. The last section provides quantitative evidence on the explosion of both the number of patent applications and their size, raising an important workload issue.

7.1. Academic Patenting

Universities are new actors in the patenting arena. Their patenting activity, which increased substantially since the early 1980s in the US and the early 1990s in Europe, is the result of two events: the emergence of new technologies (especially biotechnologies for which the distinction between basic and applied research is difficult to draw) and the adoption of the Bayh–Dole Act in the USA in 1980. The Bayh–Dole Act gave universities greater incentives to commercialize technology: 'The act allowed universities to patent the results of federally-funded research and license the resulting technology to businesses and other entities' (Joint Economic Committee US Congress 1999: 31).

The Bayh–Dole Act and the following increase in US universities' patent filings have substantially influenced EU countries' approach to the issue of patenting the results of academic research. Since the mid-1990s a considerable number of European countries have adopted similar 'US-style' legislations (e.g., the UK National Health Service circular of 1998; Germany in 1998; and Belgium in 1999 with the Decree on Education). At the same time there has been a trend towards the repeal of the professor's privilege (*Hochschullehrerprivileg*: which means that professors-inventors had the right to patent their own inventions) in Germany (2001); Austria (2002); and Denmark (2000). Other European countries have, however, adopted different legislations. Italy has introduced the professor's privilege in 2001 (*Legge Finanziaria*) and Sweden has maintained it. Beside the regulatory changes that took place, several regional governments have increasingly provided financial support to their local universities in order to support the development or the creation of the nascent technology transfer offices.

Some observers argue that the rise in academic patenting seems to have occurred irrespectively of new legislations in the US or in several European countries (see for instance Mowery et al. (2002) for the US and Meyer (2002) for Finland). Be it the result of legislative changes or not, one must acknowledge that there has been a considerable increase in the role of academic patenting.

The relative importance of university invented patents is difficult to measure, because a majority of academic inventions are filed by individuals or firms, and not by the universities (see for instance Meyer (2003) and Saragossi and van Pottelsberghe de la Potterie (2003)). Figure 7.1 provides the total number of patents filed at the EPO by universities of selected European

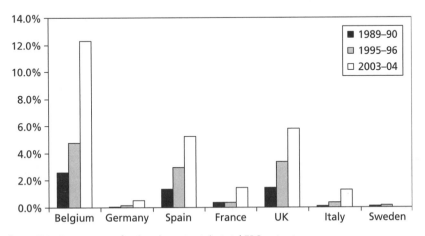

Figure 7.1. Percentage of university patents in total EPO patents

Notes: Proportion in which universities are involved as assignee in total EPO applications, by country of residence of assignees.

Source: data are provided by Steunpunt O&O Statistieken (Flemish Community, Belgium) and that the data analysis builds on the sector allocation method described in Van Looy B. et al. (2006b).

countries. It only shows a lower bound estimate of the patenting behaviour of public research institutions, because there is no information available for the patents invented by professors or academic researchers, but applied by third institutions. The surge in academic patenting is much more important in countries like Belgium (up to 12 percent of all patents filed by Belgian assignees at the EPO), Spain, and the UK. The relatively low figures for Germany, Italy, and Sweden do not mean that academics are less productive in these countries, but rather reflect different legal frameworks. Germany has only repealed the professors' privilege status in 2001, whereas it is still enforced in Italy and Sweden (as of May 2006). In these countries, academic inventions are 'managed' by the professors, not the university. For France the share is biased downward, as an important share of academic inventions are performed and applied by the CNRS, a public research centre, which is the seventh largest patentee in France, but which is not taken into account in Figure 7.1.

An alternative way to measure the importance of public research in total patenting is to search for the patents that have been filed at the EPO by the assignees with an identification that includes the words institute, university, or centre. Such exercise has been performed by Sapsalis and van Pottelsberghe de la Potterie (2006b) with EPO patents. The authors show that from less than 300 patents filed in 1981, public research institutions nowadays file about 4500 patents per year. The share of public research institution in the total number of EPO filings has jumped from about 0.5 percent of all applications in the early 1980s to a more significant 3.5 percent in the mid-2000s. A very similar trend for the relative importance of academic patents has been observed at the USPTO by Mowery and Sampat (2004). This evolution towards a more active role of universities is especially witnessed in science-based fields, such as nanotech or life sciences, where academic contributions account for up to half of the patenting activity in certain areas (see e.g., Schmoch 2004).

The justification of these policies, as well as of regional governments' support to academic patenting, finds its roots into the belief that academic patenting fosters the rate of technology transfer from the academic sector to the business sector. In addition, academic patents would increase the probability to have local (or regional) knowledge transfer and hence potential job creation, through entrepreneurial activities. From the universities' viewpoint, a sound management of their IPR would further lead to substantial financial return.

Several scholars have, however, expressed serious concerns regarding the arrival of universities in the patent arena, based on two grounds. The first one is that fostering academic patents might lead to a lower 'quality' of these patents (see, for instance, Henderson et al. 1995, 1998). However, no strong quantitative evidence is available so far to infer a negative impact (see Sampat et al. 2003). The learning experience seems, however, to play a role, as shown by Mowery and Ziedonis (2002), for US academic patents: 'Inexperienced academic patentees appear to have obtained patents that proved to be less significant

Box 7.1. Academic patenting in Belgium

Academic patenting has substantially increased over the 1990s, as illustrated in Figure 7.2. Although still with relatively small size patent portfolio, six large Belgian universities in Belgium have drastically changed their propensity to patent academic inventions. The most 'engaged' university is the K.U. Leuven, which started its valorisation much earlier than the other universities.

The relatively small number of patent can be explained by the exclusive focus on effective value creation, induced by budget constraints, either through the creation of spin-offs, or

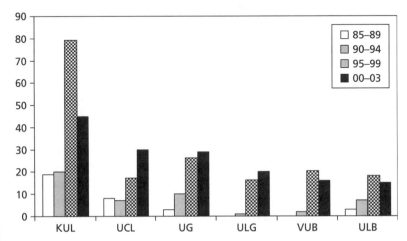

Figure 7.2. Evolution of the number of EPO patents applied by the main Belgian universities, 1985–2003

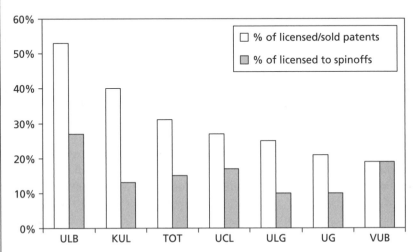

Figure 7.3. Percentage of licensed or sold patents: total and to academic spinoffs (2000–03)

through the licensing to large firms. Contrarily to the business sector, universities do not have the resources to enter into the various strategic patenting behaviours described in Chapter 4. This is illustrated by the relatively high share of licensing described in Figure 7.3, which range from 20 to 50 percent of the patent portfolio, a much larger share than the one generally observed in the business sector.

In addition, an impressive share of academic patents are actually licensed to academic spin-offs, which is also much less frequently observed in the business sector. For instance, from one spin-off in 1994, the Université Libre de Bruxelles (ULB) has increased its entrepreneurial activities to about 21 spin-offs in 2005. This entrepreneurial orientation, actively supported by regional governments, is justified by the expected job creation it may induce. The modest but fast creation of highly skilled employees by the start-ups has jumped from about ten employees in 1995 to 210 in 2005.

(Sapsalis and van Pottelsberghe de la Potterie 2006b;
Mathieu et al. forthcoming)

(in terms of the rate and breadth of their subsequent citations)'. The issue in this case is more a matter of management competencies of technology transfer offices and will naturally improve over time.

The second concern is that the incentives created by the fostering of academic patenting might shift the academic research towards more 'applied' research, away from basic (or fundamental) research, with detrimental effect on long-term growth (see, for instance, Dasgupta and David 1994; Heller and Eisenberg 1998; Nelson 2004). There is little, although increasing, empirical evidence on this issue, which seems to suggest that patents and scientific publications rather evolve within a virtuous cycle.

First, patenting does not seem to decrease the scientific production of academic researchers. Breschi et al. (forthcoming) clearly show that Italian university professors involved in at least one patent application at the EPO publish more and better quality papers than their colleagues with no patents, and increase their productivity after patenting. Zucker and Darby (1996) also find that prominent US researchers in biotechnology appeared to have outstanding research records, even after their involvement in patenting. This result is further validated by Azoulay et al. (2005) who find that scientists' patent are positively related to subsequent publication rates. An in-depth case study of the K.U. Leuven, performed by van Looy et al. (2006a), emphasizes that the publication behaviour of academic inventors differs from their colleagues (non-inventors) working within similar fields of research. Inventors publish significantly more. The only negative insight is provided by Murray and Stern (2005), who show that the citation rate of 169 articles that were published and then patented falls by 9 to 17 percent after the patent has been granted, witnessing a drop in scientific research.

Second, a high scientific performance in terms of publications and citations turns out to increase the probability of filing a patent. Breschi et al. (2006)

provide evidence with a quantitative analysis of Italian professors: 'more productive scientists are more likely to become academic inventors, to no detriment of their orientation towards basic research'. Sapsalis and van Pottelsberghe de la Potterie (2006a, b) and Sapsalis et al. (forthcoming) obtain complementary results, based on Belgian academic patents, underlining the importance of scientific publications for the patenting process of academic inventions. They first show that the value distributions of corporate and academic patents have similarly skewed distributions. Second, the researchers with a high scientific profile in terms of publications crystallise their scientific expertise—or tacit knowledge—into patents with an above than average value.

In other words, very few empirical analyses so far, to the best of our knowledge, seem to validate the concerns about a potentially detrimental effect of academic patenting on the quality of academic research. The burgeoning empirical evidence rather underlines a positive and mutually reinforcing role of academic patents and scientific publications. Even if the anti-commons hypothesis is validated by very few authors, it must be kept in mind that granted patents may enhance incentives for further research and development efforts, especially in the business sector, or witness an effective commercialization effort that is much more risky and expensive than fundamental research.

According to Nelson (2004), the increasing privatization of the scientific commons induces a risk to see an important share of future scientific knowledge becoming private property and hence falling outside the public domain. Universities should maintain their commitment to the free flow of knowledge to serve the global public interest and pursue their scientific and technological achievements (Nelson et al. 2004). Indeed, patents may lead to the blockage of some research fields, especially within the cumulative process of fundamental research activities. Nelson puts forwards several suggestions for patent policy that might avoid a sharp reduction of the 'scientific commons', while maintaining the patenting of academic inventions; these suggestions may be analysed in the light of Chapters 3 and 5:

- special care should be devoted to avoid the granting of patents on discoveries that largely are natural phenomena;
- the interpretation of the meaning of 'utility' or usefulness condition for the granting of a patent application should be more rigorous;
- the granting of patents with a too broad scope of protection should be avoided;
- a clear and explicit research exemption should be built, especially for basic research projects (despite the problem of delineating basic research): universities would be immune from prosecution for using patented materials in research if (1) those materials were not available on reasonable terms and if (2) the university agrees not to patent anything that comes out of the research.

This last suggestion on 'research exemption' deserves further consideration. The issue of whether or not academic research, or R&D activities in general, should benefit from a statutory research exemption is widely debated nowadays, especially since the famous *Madey v. Duke* decision (307 F 3d 1351, Fed. Cir. 2002) in the United States, which substantially narrowed the scope of research exemption. The case has been extensively commented because it concerns research activities performed in a university. No consensus has been reached so far. A number of countries with no exemption for research activities have expressed concerns that patent law has the potential to hinder scientific research. The crux of the problem is to understand whether or not research activities relying on (i.e., using, testing, or aiming at improving) a patented invention fall outside the scope of the patent protection and hence are not considered as infringing activities. Dent et al. (2006) provide several arguments for and against research exemptions.

On the opposing side, the argument is that patents do not prohibit research on the invention, they merely add to the cost of doing research. The appropriate level of investment incentives would require researchers to pay the full cost of any input they use. If this incentive disappears, the rate of invention may decrease. The arguments of the advocates of research exemptions are based on the potential adverse effects of patents, such as the heavy transaction costs induced by multiple licensing agreements and the fundamental uncertainty induced by basic research inventions. The main advantages induced by research exemptions are that they stimulate extensions or improvements to the original inventions, they open the door towards different technological trajectories, they generally contribute to a wider expansion and diffusion of knowledge (see the static vs. dynamic equilibrium of patents, in Chapter 3).

Europe is somewhat more favourable to research exemption than the USA. The Community Patent Convention (CPC) clearly allows for a research exemption, but its interpretation widely differs across the EPC member states. Article 27(b) states that: 'the rights conferred by a Community patent shall not extend to . . . (b) acts done for experimental purposes relating to the subject-matter of the patented invention'. According to Dent et al. (2006), a very liberal approach has been taken by the German Constitutional Court, which concluded that patent owners had to 'accept such limitations on their rights in view of the development of the state of the art and the public interest'.

The question of whether the patent system hinders research activities in general and academic research in particular is increasingly being investigated. However, little evidence on the issue, a few case studies excepted, is so far available. The most comprehensive insight is probably provided by the two surveys of US biotech researchers performed by Walsh et al. (2003, 2006). There seems to be no or little breakdown or restricted access to research tools because firms and universities have been able to develop 'working solutions' that allow the researchers to proceed. In addition, academic researchers tend to disregard existing patents (only 5 percent of 381 respondents check regularly for patents on knowledge or material inputs) and adopt a rational forbearance of potential infringement. As their research objective is mostly for non-commercial uses, they do not feel threatened. Similar results were obtained, though with a smaller sample, with German scientists. Strauss (OECD 2002: 47) found that patents on research tools have no discernible effect on the cost and past of research, because it is difficult to detect infringement and the researchers are unaware of the legal implications of using patented research tools.

Walsh et al. (2006) also found that 19 percent of researchers did not receive the last requested research input (out of an average of seven requests to other academics and two requests to the industry). The reasons for which scientists do not provide materials include the commercial orientation of their research, the funding by the industry, scientific competition and their number of publications, or scientific eminence.

It can legitimately be assumed that universities are unlikely to become the target of patent infringement suits because the damages would be too small to justify the cost of litigation (Rowe forthcoming). In addition, an intensification of such litigations against the academic sector may seriously damage the perception of the effectiveness of the patent system in stimulating innovation. Firms have little to gain and a reputation to lose. On the other hand universities are increasingly relying on the patent system, and do enforce their proprietary knowledge, inducing a higher probability of patent litigations against their research activities, either from the business sector, or from the academic sector. The fact that the research performed by universities is increasingly being funded by the business sector may well intensify the debate on research exemption in the future.

7.2. **Comparative Analysis of Costs**

One of the most sensitive and long lasting bones of contention regarding the European patent system is related to the high cost of patenting. The cost of patenting can be divided into four main categories associated with the granting process (see Roland Berger 2005; van Pottelsberghe de la Potterie and François 2006):

- *Process costs* are composed of procedural fees (such as filing fees, search fees, examination fees, country designation fees, claim-based fees, and country specific validation fees for the patent granted by the EPO).

- *Translation costs* consist in the translation services often provided by patent attorneys. These costs occur mainly once a patent is granted, and depend on the size of the patent (the number of pages) and on the strategy regarding the geographical scope for protection. The larger the number of countries, the higher is the cost induced by translation services. Measuring these costs is not straightforward as they include translation and transaction (intermediation by patent attorneys) costs.

- *External expenses* consist in the services costs associated with the writing of the patent and the filing to a patent office. While large firms often have their own IP department, with officially accredited patent attorneys, small firms must always rely on the services provided by legal advisors and accredited patent attorneys. The costs of these services include the expenses associated with all actions implemented for a patent: filings, payment of fees, monitoring translations, and procedural actions (time spent in oral or written communication with the patent office).

- *Maintaining costs* are composed of the required renewal fees to keep the patent valid during a maximum period of 20 years. They are due each year at the JPO and at the national patent offices of the EPC member countries or after a certain period at the USPTO; they are generally increasing progressively over time. Once a patent is granted by the EPO, it must be validated and put in force in each desired national patent office of the EPC member countries. The renewal fees vary significantly across countries.

Calculating patent costs is far from being straightforward, as several components are not easy to quantify and depend on the applicant's strategy for filing a patent (e.g., the number of claims, the number of pages, the route, the quality of external services, the desired speed of the granting process, and the targeted geographical scope for protection). Larger patents (i.e., with more claims and/or more pages) and patents that applicants are seeking to have protected in a large number of EPC member states are more expensive in terms of both process cost and external cost. The cost is further linked to the delay of the procedure (especially if a significant number of written communications take place between the patent attorney and the patent office, as illustrated in Chapter 6) and to the desired speed of the granting process. Whereas administrative or procedural expenses are easy to quantify (official fees), the external services and translation costs must be approximated, as they depend more on the quality of the services provided by attorneys.

Two approaches can be used to assess the cost of patenting. The first one is to directly ask companies the cost they incur for their patent applications. It provides a direct evaluation of the external expenses effectively supported by the sample surveyed. However, it depends on the sample selection (i.e., the firms that can afford the system, which are mainly large firms). The second approach consists more in a simulation, based on the known parameters, such as fees. As the cost of patents depends on the targeted number of countries for protection, on the number of claims and on the external expenses, various combinations can be used. The first approach, which has been performed for the EPO by the Roland Berger (2005) survey of applicants, is illustrated in Table 7.1.

Table 7.1. Average cost of patents for ten years of protection

	Euro-PCT (RB-PCT-8)	Euro-Direct (RB-ED-6)
Hypotheses:		
Designated states	8	6
Number of claims	15	6
EPO fees	6 300	4 400
Prof. representation	12 150	9 630
Validation:		
Attorney	4 000	2 850
Translation costs	7 400	3 400
Total	12 100	6 650
Renewal fees:		
Fees	9 200	5 600
Attorneys	6 800	5 300
Total	16 000	10 900
Total attorneys and prof. rep.	22 100	17 780
Grand total	46 550	31 580

Source: The Roland Berger (2005) survey sponsored by the EPO.

The survey distinguished between the applicants going through the PCT route and those going through the Euro-direct route (direct or second filing at the EPO, see Chapter 6). PCT applications effectively designate height EPC member states for protection and include about 15 claims. Euro-direct applications generally target six countries and include only six claims on average. For ten years of protection, the total cost of PCT applications is of €46,000, whereas the Euro-direct application is slightly above €30,000. The expenses induced by attorneys and professional representations account for about half of these costs. Translation expenses range from 11 to 15 percent of total costs, the rest corresponding to the fees (more than 30 percent of total costs).

The second approach has been followed by van Pottelsberghe de la Potterie and François (2006). The cost of patenting was simulated for the 'average patent' filing (in terms of the number of claims) at the three offices (EPO, JPO, and USPTO). The costs of priority filing in a national patent office were not taken into account. Similarly, the costs induced by the PCT route were not taken into account. In short, the PCT route provides more time to the applicants to assess the real value of their patent and slightly increases the total cost of a patent filing. It is therefore important to keep in mind that not taking into account the costs associated with national priority filings and PCT filings means that the cost simulations that follow should be considered as a lower bound of the real cost that an applicant would have to bear. The simulated costs are presented in Table 7.2.

Table 7.2. Simulated cost structure (in EURO) of direct patent filings and maintenance at the EPO, the USPTO, and the JPO (2003/04)

	EPO-3[1]	EPO-13[1]	USPTO	JPO
Hypotheses:				
Median number of claims	18	18	23	7
Time to grant (number of months)	44	44	27	31
Designated countries for protection	3	13	1	1
Number of translations[2]	2	8	0	0
Procedural costs:				
without translation	4 670	6 575	1 856	1 541
with translations[2]	8 070	20 175	1 856	1 541
External services cost[3]	12 500	19 500	8 000	4 000
After grant:				
Maintaining costs 10 years (fees)	2 975	16 597	2 269	2 193
Maintaining costs 20 years (fees)	22 658	89 508	4 701	11 800
Total filing process until validation	20 570	39 675	9 856	5 541
Total 10 years	23 545	56 272	12 125	7 734
Total 20 years	43 228	129 183	14 556	17 341

Notes: [1]The three EPC member countries that are the most frequently designated for protection are also the largest European countries: Germany, Great Britain, and France. According to the EPO annual report of 2003, the 13 countries that are designated for protection by more than 60 percent of the patent applications are: Germany, Great Britain, France, Italy, Spain, the Netherlands, Sweden, Switzerland, Belgium, Austria, Denmark, Finland, and Ireland. [2]Only 8 translations would be required for an effective protection in 13 countries as some countries accept applications written in English or share a common language with other countries (the Netherlands, Belgium, Switzerland). It is assumed that translation costs are of €1700 per language. This amount includes the translation and attorneys' intermediation. [3]There is no existing comparisons of external services costs in the US, Japan, and Europe. The Roland Berger (2005) survey provides a reliable estimate for applications at the EPO. We assume a base of 8000 for a patents and £1500 per designated states (for the EPO). As the patents applied at the JPO are much smaller (7 claims against 18), we assumed half the base cost (i.e., €4000).

Source: Adapted from van Pottelsberghe and François (2006).

Two different scenarios have been used for the EPO. In the first one the patent only aims at designating the three largest European countries (EPO-3: Germany, France, and the UK). In the second scenario, it is assumed that the assignee designates 13 countries (the countries that are actually designated by at least 60 percent of the patents filed at the EPO during the examination stage, see section 6.1). The first rows of Table 7.2 describe the hypotheses that have been used. It is implicitly assumed that a patent is granted and then renewed for ten or 20 years (in other words, if a patent is withdrawn before the grant, or refused by the patent office, the cost would be lower). This hypothesis is made in order to reflect the costs that an applicant must be ready to bear when starting an application. The number of claims corresponds to the average number of claims observed in each patent office in 2004 (see section 7.5 for an analysis of the number of claims).

Procedural costs consist of filing, search, examination, designation (exclusively for the EPO), grant, validation, and administrative fees. A total of €4670 and €6575 is due to have a patent granted at the EPO when three and 13 countries are designated, respectively. Contrary to the European patent, filing a patent at the USPTO or the JPO does not need to be translated (except for foreign applications written in another language than English or Japanese; respectively) and no validation fees are required. For the European patent, translation costs, national taxes, and validation charges have to be accounted for once the patent is granted. After the grant, the applicants have to pay validation fees to each national office in order to put the patent into force.

When a patent is granted by the EPO, it transforms itself into a bundle of national patents (in all or a selection of the countries that were designated by the applicant). National laws stipulate that a patent written in a foreign language has no effect, legally speaking. Applicants need therefore to translate their patent documents. These translation costs are considerable and mandatory if the patentee wants a protection in every European country. For the 13 European countries eight translations are required and would cost the applicant about €13,600.

A patent application is most often performed with the help of patent attorneys who guide the assignees through the whole procedure. A tentative estimate of external service costs is displayed in Table 7.2, it should be cautiously interpreted as there is no existing reliable and comparable evaluations of these costs in the US, Japan, and Europe. The amounts presented here must be taken as a crude approximation, because they may vary significantly across firms, industries and countries. In addition, a significant share of these external costs can be considered as internal expenses for large firms (i.e., depending on the number of staff in the IP department), which makes them further complicated to measure. So far, the Roland Berger (2005) survey implemented for the EPO is the most reliable source of information on external expenses, especially for patent applications at the EPO.

For each country and each year the renewal fees must be also added. For the 13 countries and for a period of ten years, the maintenance fees add up to €16,597 (€2975 for three countries). This amount is more than four time as high for a period of 20 years (€89,508 for 13 countries). For the US and Japanese market, the ten years' protection costs vary between €2000 and €2500.

At the USPTO maintaining fees are due 3.5, 7.5 and 11.5 years after first filing, which means that a patent maintained for 20 years does not cost more than a patent maintained for 12 years. A patent maintained for 20 years at the USPTO would cost about €14,500 for a large firm (SMEs only pay 50 percent of the fees). The fees in Japan depend heavily on the number of claims: the larger the number of claims, the more expensive the patent. Procedural costs are quite low for a patent application at the JPO. An average Japanese patent (with seven claims) that is renewed for 20 years would cost about €17,000.

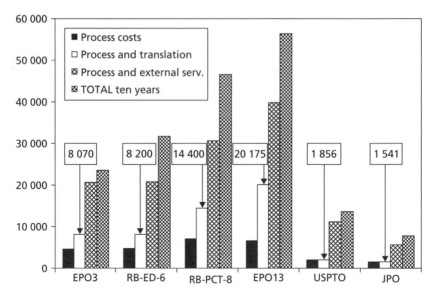

Figure 7.4. Comparative (€) cost structure of patent application and maintenance for ten years (2003–04)

Source: Adapted from Roland Berger (2005) and van Pottelsberghe and François (2006) (cf. Tables 7.1 and 7.2).

Figure 7.4 illustrates the cost differences across the three patent systems, for a ten year enforcement period. The European patenting cost, approximated through the Roland Berger survey (2005) and the simulations of François and van Pottelsberghe de la Potterie (2006), clearly display higher costs in Europe than in the US or in Japan. This is mainly due to translation and validation costs as well as to the renewal fees that the EPO applicant has to fulfil. The larger the number of designated states for protection, the more expensive the patent is, as both renewal fees, translation costs, validation fees, and external services increase with the number of countries. A European patent that is renewed for ten years would cost between €12,000 and €56,000, depending on the number of countries targeted. A protection sought in all EPC member states would be beyond €130,000, against about €14,000 and €8000 for the US patent system and the Japanese patent system, respectively. The Japanese patent system is the least expensive for the process costs. For a 20 years protection the US system would be the least expensive (see Table 7.2).

This type of comparison has to be taken with caution because (1) it is calculated for the 'average' patent in each patent office; and (2) external services expenses are estimates and can vary substantially according to the technology and the strategy adopted by the firm. Van Pottelsberghe de la Potterie and François (2006) suggest that it might be more relevant to perform a comparison on the basis of claims, rather than on the basis patents. They show that the process and translation costs of a claim that is granted by the EPO goes from more than €400 for three designated states to more than €1000 for 13 designated states, as compared with about €80 and €220 per claim in the US and Japan, respectively. The cost structure of a patent applied only in the three largest EPC countries is very close to the one of Japan.

The patenting costs of the EU and Japan relative to the US are presented in Table 7.3. It shows that a European patent is four to ten times (depending on the number of countries that are targeted) more expensive than a US patent if process and translation costs are considered. For the total costs with up to ten years of protection, European patents would be nearly two to five times more expensive. Japanese patents are 50 percent less expensive than US patents. If the analysis focuses on the claims, the cost differences increase, as there are fewer claims on average in an EPO patent, and especially in a Japanese patent vs. a US patent. One claim, the lowest common denominator of a patent, is six to 14 times more expensive in Europe than in the US, as far as process and translation costs are concerned. The Japanese claims are at least two times more expensive than in the US.

The literature seems to suggest that what drives firms' patent behaviour is not related to the cost of patenting (see, for example, Peeters and van Pottelsberghe de la Potterie 2006a; Duguet and Kabla 1998) but rather to the strategic behaviours described in Chapter 4. In order to investigate a possible relationship between the cost of patenting and the demand for patents (i.e., the total number of patent filings) we relied on the costs that have to be supported

Table 7.3. European and Japanese patent costs relative to the US

	Process and translation	Total costs (10 years)
Relative US patent:		
EPO3	4.3	1.9
RB-ED-6	4.4	2.6
RB-PCT-8	7.8	3.8
EPO13	10.9	4.6
JPO	0.8	0.6
Relative US claims:		
EPO3	5.6	2.5
RB-ED-6	16.9	10.0
RB-PCT-8	11.9	5.9
EPO13	13.9	5.9
JPO	2.7	2.1

Note: These figures represent the cost of a European or Japanese patent (claim) divided by the cost of a US patent (claim).

Source: Computed from Roland Berger (2005) and van Pottelsberghe and François (2006), see Tables 7.1 and 7.2.

until the grant of a patent and its validation in the designated states (including procedural costs and translation costs) and we exclude external expenses (e.g., attorneys and professional representations) as they are difficult to approximate, especially within an international comparison framework. Procedural costs and translation costs correspond to the minimum expenses assignees will have to foresee when they apply for a patent.

The relationship between the cost of the patent process and the number of patent filings must take into account the market size associated with a patent. If two patent systems offer the same cost structure for the granting of a patent, and if the two regions are of different size, the assignee would have a clear preference for the region with the largest market size. In other words, the largest region would offer protection for each unit of the market (e.g., a consumer) at a lower cost. This approach requires computing the cost per claim per capita, or the 3C-index suggested by van Pottelsberghe de la Potterie and François (2006).

Figure 7.5 takes into account the market size associated with the three patent offices. It shows the relationship between the process cost per claim per million capita and the total number of claims that have been filed in the patent offices. A negative and linear relationship clearly appears between the price of patents and the demand for patents (as measured with the average number of claims in all patent filings). The US having a large market, its relatively low cost per claim becomes even lower 'per capita', about €0.3 per million inhabitants, which explains its high attractiveness. In Japan the cost per claim per million capita is

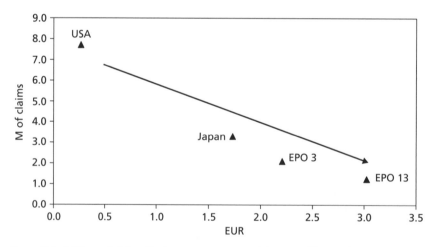

Figure 7.5. Millions of claims filed v. process cost per claim per million capita

Note: The x-axis shows the cost per claim per million capita, expressed in €, and includes the procedural fees and translation costs. The y-axis shows the number of claims filed (millions) in each patent office (cf. Table 7.2 for the list of countries included in EPO3 and EPO13).

Source: Van Pottelsberghe and François (2006) (cf. Table 7.2).

Table 7.4. EPO, USPTO, and JPO: basic figures (2004)

	EPO[1]	USPTO	JPO
Total patent filings	123 706	356 943	423 081
Total patent granted	58 730	164 293	124 192
Geographical origins of patent filings (%):			
USA	26	53	5
Japan	17	18	87
EPC states	49	15	5
Others	8	14	2

Note: [1]European applications filed and Euro-PCT applications entering the regional phase (2004).

Sources: Annual Reports of the three offices and the Trilateral Statistical Report (2004).

about €1.7, whereas it is about €2.2 and €3 for the three and 13 most frequently designated states in Europe, respectively. The higher the patenting cost per claim per capita, the lower the demand for patents. The 3C-index shows that a claim applied in 13 European states is about ten times more expensive than in the US, and there are six to seven times fewer claim filings in Europe than in the US.

This relationship between price and the demand for patent may partly explain the differences between the three patent offices in terms of patent filings. Table 7.4 shows that there are far fewer applications at the EPO (123,706 direct applications and PCT filings in the regional phase in 2004) than at the USPTO or the JPO, which received more than 355,000 and 423,000 filings respectively, in 2004. This difference may be explained by three main factors.

First, the filings at the EPO are mostly second filings—i.e., they are subsequent applications of existing priority filings, as clearly described in Chapter 6. Contrarily to the European system, the numbers of domestic filings at the JPO and the USPTO are approximately equivalent to the number of domestic priority filings (the first time a patent is filed for an invention). In most EPC contracting states the priority filing is generally applied first at the National Patent Office (although a substantial proportion is directly filed at the EPO). After the national priority filing assignees have one year to extend their application internationally. Due to the differences in behaviour of the applicants from different countries (only a share of national priority applications are filed at the EPO), comparisons of the number of applications at the three offices requires a cautious interpretation. Second, patent filings at the three offices are different. In 2004, the average number of claims in a Japanese patent was about eight, whereas it was 24 for the USPTO and 18 for the EPO. In other words, a given invention would be protected through more patents in Japan (and fewer claims per patent) than in the United States and the European Union. Third, the European patent system might be less attractive than the other two patent systems, due to its cost and its complexity, as illustrated in Figure 7.5.

Table 7.4 illustrates a second important difference. Foreign filings are much more important at the USPTO and the EPO than at the JPO. The fact that the US and Europe similarly attract an impressive number of foreign patent filings (about half of the filings come from abroad) reduces to some extent the statistical bias mentioned here. There are nearly 170,000 patent filings at the USPTO by non-US applicants, and about 62,000 filings at the EPO by non-European residents. The picture is similar with the patents filed by Japanese residents. The figures in Table 7.4 suggest that only about 21,000 filings were applied at the EPO by Japanese firms, against 64,200 filings at the USPTO.

The three indicators of attractiveness (total patent filings, filings from abroad, and filings by Japanese assignees) suggest that the US market for technology is three times as attractive as the European market for technology. One explanation most probably finds its root in the cost factor illustrated in Figure 7.4. Another explanation might be related to the perception that there is a lack of a homogenous market for technology in Europe. One of the major differences between the European patent system on the one hand and the US and Japanese patent systems on the other hand, is that once a patent is granted it must be validated and put in force in each EPC member state. A patent that is granted by the EPO is not automatically valid all over Europe, and validations translate into much larger costs (see Figure 7.4).

It is important to keep in mind that what makes a Europe-wide patent more expensive is mainly the translation costs (including the intermediations with patent attorneys) and the renewal fees in each national patent office that occur just after the grant of the patent by the EPO. They seem to affect the demand side of patents, through an implicit cost-based self-selection process, which is implicitly more favourable to large firms. These costs do not correspond to the quality of the search and examination process of the EPO, and hence do not witness a more rigorous examination process. This latter dimension is analysed in the next section.

7.3. **Comparative Analysis of Operations**

Although the broad objectives of the three patent offices are similar, the way they are implemented and their effectiveness in fostering innovation may differ. The previous section leads us to wonder whether the much higher cost of patenting in Europe is associated with a higher quality service offered by the EPO? This section underlines some important differences between the three patent offices, regarding the procedural steps and their potential impact on quality. These differences, which might be considered as rough indicators of the rigor of a granting process, are illustrated by the workload, the speed of the

Table 7.5. Analysis of operations at the three regional offices (2004)

	EPO	USPTO	JPO
Number of examiners	3 365	3681	1243
Procedure pendency (in months, 2003):			
Search	12	27	–
Examination	38	–	31
Total patent filings	123 706	356 943	423 081
Total patent granted	58 730	164 293	124 192
Total number of claims filed (million)	2.2	8.6	3.0
Filings per examiner	36.8	97.0	166.4*
Grants per examiner	17.5	44.6	48.8*
Claims filed per examiner	654	2 336	1 180*
Claims granted by examiners	314	1 147	354*
Grant rate[1]	59% (QW: 67%)	64% (QW: 87 to 97%)	50% (QW: 64%)

Notes: *In Japan the search process is outsourced to an external organization composed of about 1300 employees, which would bias a 'per examiner' comparison. In this table we added the total number of examiners (1243) to the approximated total number of employees devoted to the search process (about 1,300); [1]Quillen and Webster (2001) and Quillen et al. (2002) put forward grant rates for the period 1995 to 1999 for the EPO and the JPO, and for the period 1993 to 1998 for the USPTO. The authors show that the USPTO grant rate (allowances divided by total disposals, i.e., the sum of allowances and abandonments), corrected for the continuous applications, ranges from 87% to 97%, depending on the extent to which prosecution of abandoned applications was continued in refilled applications.

Sources: Trilateral Statistical Report (2004); Quillen and Webster (2001); and Quillen *et al.* (2002).

process, the grant rate and the patentability of specific subject matters. Table 7.5 summarizes some of the differences across the three patent offices.

With approximately the same number of examiners, the USPTO tackles about three times more patent applications (or patent granted) than the EPO. If the number of claims is considered, it appears that the USPTO supports a workload about four times heavier than that of the EPO. At the EPO (v. the USPTO), 654 (v. 2336) claims are filed per examiner, and 314 (v. 1147) are granted each year. These figures suggest that both the incoming workload of examiners (number of claims filed per examiner) and their output (number of claims granted per examiner) is three to four times higher at the USPTO than at the EPO. One explanation of this significant difference might be due to the time spent by an examiner on each patent.

The length of the procedure is an important factor for applicants as it shows the time needed to get a patent granted. The EPO has the greatest average delay (49 months) compared to the JPO (31 months) and the USPTO (27 months). One must be cautious with these figures because both the EPO and the JPO have to wait for an answer from the applicant during the examination process (see Chapter 6). After the search process and the publication of the application, the EPO and the JPO have to wait for a request for examination by the applicant; it is allowed to wait for up to three years to answer to the JPO and six months to the EPO (cf. Chapter 6). It is however well known that the EPO takes more time for the search and examination processes than the USPTO and the JPO.

Examination is performed faster at the JPO, partly because the search is done by the IPCC and that a Japanese patent application is composed of, on average, only seven claims, which obviously lightens the search and examination process per patent. Regarding the USPTO, Lemley (2001: footnote 5) reports that 'there are strong structural and psychological pressures on examiners to issue patents rather than rejecting applications, no matter how weak the alleged invention seems'. Examiners at the USPTO have only about 18 hours per patent for the whole examination process.

A slower process means that examiners spend more time on a patent application. Assuming similar analytical skills, it can logically be inferred that the EPO examiners' decisions are based on a broader knowledge of the prior art and a deeper analysis of the patented invention, that would lead to a higher quality of the granted patents (i.e., a higher refusal rate).

The substantive examination is a crucial step in the process. When an examiner intends to grant a patent or if a patent cannot be granted as such, the information is communicated to the applicant. The latter may then make amendments to the application, generally in the number and content of claims, after which examination is resumed (cf. Chapter 6 for a description of the interactions between applicants and the patent offices). This procedural step is iterated as long as the applicant continues to make appropriate amendments. Then, the patent is either granted or the application is finally refused (or withdrawn by the applicant). An applicant may always withdraw the application, at any time before the decision of the patent office.

The examination process at the EPO is further reinforced by the post-grant opposition system that allows third parties to challenge the patent applications before they are effectively issued (at the EPO an opposition can be filed for nine months after the grant of a patent). An opposition process is much less costly than litigations, and induces a fast proceeding of the cases, which in turn reduces uncertainty regarding the patentability of the invention. One could consider the opposition process as a clear upgrade in the rigor of a patent system, as it generally adds useful information (new prior art) about the invention and its patentability, and would clearly contribute to reduce the grant rate of a patent office. About 7 percent of all the patents granted by the EPO are opposed each year.

The JPO has the lowest grant rate (50 percent), making it the less 'applicant-friendly' office compared to the two others. The USPTO's grant rate (64 percent) is slightly higher than the one of the EPO's (59 percent). However, according to Quillen and Webster (2001) and Quillen et al. (2002) this rate is biased because it does not include the continuous applications in its calculation: the USPTO's grant rate should be corrected and would fluctuate between

87 and 97 percent, making the American office the most 'applicant-friendly'. There are three types of continuing patent applications available in the US pursuant to the patent statutes: continuations, continuations in part (CIPs), and divisionals. Continuation and CIP applications permit patent assignees to refile their patent applications and hence restart the examination process with a newly filed patent application claiming the benefit of an earlier filing date. The prior application is normally abandoned after the second application is filed (see Quillen and Webster 2001: 4). This corrected grant rate should be compared with the grant rates of 67 and 64 percent for the EPO and the JPO (for the period 1995–99), respectively. These figures tend to show that the JPO and the EPO have adopted a much higher level of rigor than the USPTO.

On average, only about 4 to 5 percent of patent applications at the EPO are refused by examiners. As 60 to 65 percent of patents are granted, the remaining applications are withdrawn by the applicants, about 30 percent of total filings. A substantial proportion of these withdrawals takes place after a communication from the EPO (i.e., search report or communications during the examination process) and hence could be considered as 'induced withdrawals'. Lazaridis and van Pottelsberghe (2006) show that 54 percent of all withdrawals can be considered as being induced by the work of examiners. 'Induced withdrawals' and refusals occur for up to 23 percent of all applications at the EPO, which is an additional indicator of the rigour of the EPO process.

Domains of patentability can vary depending of the region. Historically, patentable subject matter has been restricted first to mechanic devices and their manufacturing processes and then extended to chemicals and pharmaceuticals. Subject matter now covers biotechnological inventions and software in most countries, but changes have not occurred everywhere at the same pace and differences remain in several dimensions (Guellec and Martinez 2003). One of the first differences across jurisdictions in this regard lies on the legal definition of patentability, and in particular what is considered a technical invention. Patent laws also differ as regards exclusions from patentability. In the US, only laws of nature, natural phenomena, and abstract ideas have been traditionally excluded from patentability. In Japan the exclusion extends to medical methods, laws of nature and discoveries. In contrast, apart from medical methods, the EPC excludes a long list of items if they are claimed as such. This list includes scientific theories, mathematical methods, aesthetic creations, schemes, rules, and methods for performing mental acts, playing games or doing business, programs for computers, and presentations of information. The long list of domains that are excluded from patentability is probably an additional factor underlying the smaller number of patent applications at the EPO.

In a nutshell, the USPTO has far fewer restrictions on subject matters, is faster, is more 'applicant friendly' and grants a much higher number of patents per examiner than the EPO. The slower speed of the EPO and the lower apparent 'productivity' of patent examiners can be explained by the greater rigor of the EPO, and a higher perceived quality of the granting process. This rigor requires longer search and examination processes and induces a more stringent selection rate—and so lower grant rate. The lower the grant rate, the more the process can be seen as strict and severe.

This dichotomy is somewhat confirmed by Jaffe and Lerner (2004)'s assessment of the current US patent system (as opposed to the late 1970s and early 1980s): 'Now that it is possible to get a patent on unoriginal ideas, many more dubious applications are being filed . . . the current system provides incentives for applicants to file frivolous patents applications, and for the patent office to grant them . . . Patents in Europe and Japan remain harder to get . . .' (pp.5–6).

7.4. **Towards a More Integrated Market for Technology?**

Several avenues are currently being investigated to lighten the burden of patent costs and the complexity of the patent system in Europe: the Community Patent, the London Protocol, and European Patent Litigation Agreement (EPLA). The high cost of patenting is partly due to the failure, so far (as of May 2006), to effectively implement the European community patent or the more recent London Protocol or European Patent Litigation Agreement. The high opportunity cost of this failure, for firms and society at large, is perceived as one important factor hindering innovation in Europe, which explains why the European Union' commissioner of Directorate General Internal Market, Mr. McCreevy, wants to make 'one final effort' to resolve the issue (*Financial Times*, 16 January 2006: 'One final effort to create a low-cost EU patent'). An EU-wide consultation process has been launched in early 2006 by the European Commission. The results have been released in July 2006, witnessing the sustained efforts to reach a more centralized and less expensive system.

As explained in section 6.1, the current European patent system is a two tier system where patent can be applied, granted, and enforced at the national level, or they can be applied at and granted by the EPO, but enforced at a national level. The two options lead to national patents that can only be enforced in national courts. The drawback of this system, beside the high cost induced by the translations and the national renewal fees, is the complexity of facing the jurisdictions of so many courts, and multiple litigations for a single patent, with potential opposite outcomes. Several attempts for further harmonization have failed over the past 30 years. The Luxembourg conferences on the Community Patent (in 1975, 1985, and 1989) all failed to improve the situation.

The first and most important potential progress in this respect would be to implement the Community Patent. Such a system, which has been under intensive debate for about 30 years, would have the following advantages: (1) a uniform and autonomous patent system for the whole European Union market—one unique gateway all along the life of the patent, with a unique registry; (2) a much less expensive process for patentees; and (3) a centralized court and with the defendant language being the language of proceedings. With such a system, a patentee would reach the 25 EU member countries at a cost that would be roughly similar to the current cost that has to be supported for the three largest European countries for instance. One of the most important difficulties with the Community Patent is the number of languages it has to be translated into. The current proposal is far from being attractive to the business sector, as it would automatically require the translation of claims into all 20 languages of the European Union. In addition the introduction of the Community Patent would require unanimity among the 25 EU member states, which reduces the probability of its implementation in the near future.

An alternative solution to overcome the translation issue has started in 1999, at the Paris Intergovernmental Conference, which set up a working party on translations. This working party put forward the 'London Protocol'. It basically aims to reduce translation costs. The London Protocol stipulates that 'any state having an official language in common with one of the official languages of the EPO shall dispense with the translation requirements provided for in Art. 65 of the EPC'. And 'any state having no official language in common with one of the official languages of the EPO shall dispense with translation requirements if the European patent has been granted in the official language of the EPO prescribed by that state, or translated into that language'. The translation requirements would therefore be as follows. The claims of a patent would only have to be translated into the three official languages: English, French, and German. The whole document would have to be translated into a maximum of three languages, depending on the choice of the signatory states. It is only under litigation proceedings that any party can ask the applicant to provide a full translation of the document.

The London Protocol would obviously drastically reduce translation costs. It must be ratified by only eight countries, including England, France, and Germany, all of whom must be included, before it can come into force. So far, as of May 2006, six EPC member states have ratified the London Protocol (Germany, the United Kingdom, Slovenia, Iceland, Latvia, and Monaco). The issue is highly debated in Denmark, the Netherlands, Switzerland, Sweden, and France, the latter being a mandatory signatory for the effective implementation of the London Protocol.

The European Patent Litigation Agreement (EPLA) is a third solution aiming at simplifying the European patent system. As for the London Protocol, it is a proposal put forward by a working party set up at the 1999 Paris

---

Box 7.2. The RIM case, part 3: *RIM v. InPro*, a complex market for technology

---

The *RIM v. NTP* case, already introduced in Box 4.2 and Box 6.2 illustrates some claim drafting practices adopted by NTP. Here we illustrate how the European patent system affects RIM's IP strategy in Europe.

InPro is a patent holding and licensing company incorporated in Luxembourg which commenced action against T-Mobile in Germany in 2005, alleging that T-Mobile was infringing its European patent by offering and implementing the Blackberry system in Germany. RIM intervened in support of T-Mobile, its customer. The case was an important one for RIM since the German company were allowed to have Blackberry services everywhere in the world.

Germany is unique in Europe for having a dual legal system regarding patent disputes. The issues of infringement and of validity are tackled in different court. The former is addressed in the District Courts, whereas the latter are addressed in the Federal Patent Court. In practical terms, it means that some case might last longer due to this bipolar system. RIM's response was to commence action in the UK, seeking revocation of the UK element of InPro's patent. According to RIM, the purpose was to convince the Düsseldorf court that there were good grounds for doubting the validity of the InPro patent.

The UK court eventually found the InPro patent to be invalid. Ironically the German case ended earlier than the UK one. It also found the InPro patent invalid and revoked it. This *RIM v. InPro Licensing Sarl* case highlights (1) the need for a single European patent court; and (2) the fact that innovators like RIM are well defended in Europe against patent trolls who try to abuse the system (the abuse seems to be witnessed by the revocation of InPro's patent). It might be worth it to compare this case with the *RIM v. NTP* case in the US (see Box 4.3).

(Moss and Grzimek 2006)

Intergovernmental Conference. It is designed to improve the enforcement of European patents, enhance legal certainty, and promote the uniform application and interpretation of European patent law. Although not tackled in section 7.2, litigation costs are an important component of the total expenses associated with a patent filing. The implementation of the EPLA would contribute to reduce the fragmentation of the European market for intellectual property (see Box 7.2). As for the London Protocol, the EPLA is optional among the 31 EPC countries, which makes it easier to implement than the European Community Patent.

THE EXPECTED ECONOMIC IMPACT OF AN IMPROVED INTEGRATION

The three aforementioned attempts to reach a more integrated market for technology would affect both the cost of patenting in Europe and the implicit geographical scope for protection. Whereas the current system consists in granting a European patent that must be validated and enforced in each national jurisdiction, the Community Patent or the EPLA would allow applicants to look for a much larger geographical area for protection. These attempts would arguably lead to a stronger and less expensive patent system, especially regarding its geographical reach. A similar tendency can be observed at the global level as well. The WTO trade related intellectual property agreements, or TRIPS, aim at extending the patent protection of developed economies towards their trade partners. In other words, the geographical dimension of the patent system is evolving over time, towards an international expansion, as clearly illustrated in the historical perspective presented in Chapter 2.

In this respect, a crucial issue is to assess whether a stronger—i.e. covering a geographically larger area—and less expensive intellectual property system would lead to more innovations. This question is of utmost importance nowadays, especially in the context of the Lisbon Agenda. It is, however, difficult to evaluate such an impact on innovation, as patent systems are more effective for some industries than for others (see Chapter 4). It is easy to find arguments against the costs induced by translation requirements. They basically are a deadweight loss: a waste for the society in general. Most of the translated patents provide no or little information, and they are generally not read. If one agrees with the idea that most scientists nowadays are able to speak a common language, English for instance, then there are few benefits induced by translations. But the Community Patent, the London Protocol, and the EPLA together aim at a strengthened European patent system. Some theoretical insights can be made about the economic impact of a more centralized system. According to Scherer (2002), the extent to which a stronger patent system may stimulate research activities depends on the extent of competition between research paths.

The issue can be illustrated through the early debate induced by William Nordhaus (1969) and Yoram Barzel (1968). Assuming that innovators maximize the present value of their expected return from process innovations, Nordhaus tried to identify the patent length that government might set to maximize the social surplus (producers and consumer surpluses minus the R&D cost). The author found that the optimal patent life varied from 1 to 34 years, depending on the cost reduction induced by the innovation and on the price elasticity of demand for the product. The easier it is to achieve cost reductions and the more price-elastic product demand is, the shorter the optimal patent life. In his model, Nordhaus implicitly assumes 'exclusive prospects' rights: the firm conducting research on a particular project is the only firm involved in this project, and it can therefore choose the level of R&D expenses that maximizes the net present value of the quasi-rents. In other words, Nordhaus claims that there is no parallel R&D project conducted by other firms.

The alternative view put forward by Barzel is that innovators compete for priority and hence temporary monopoly profits. This inter-firm competition would drive aggregate R&D costs upward until the aggregate net profit from inventive activity (netting the losses of losers and the gains of winners) would be zero on average (from the social viewpoint). Barzel's view is called the 'rent seeking' model.

Scherer (2002) shows that, (1) at the aggregate level, the rent seeking assumption (of Barzel) would lead to more R&D expenses than the exclusive prospects assumption (of Nordhaus), for a given patent system; and (2) a strengthening of the IP system would have strong stimulating effects under a situation of exclusive prospects, and a smaller, or incremental, effect under the rent-seeking assumption.

Figure 7.6, which is adapted from Scherer (2002) illustrates the two theoretical models, and the implications of weaker or stronger intellectual property regimes. The horizontal axis measures two variables: the amount spent on biopharmaceutical R&D and the discounted quasi-rents (sales less costs of production and distribution) resulting from that R&D. The vertical axis measures the number of new biological entities (NBEs) entering the market, as a result of R&D. The solid line C (RD) is the invention possibilities function, which shows that the number of NBEs depends on the amount of R&D done. It exhibits diminishing returns in the R&D–NBE relationship (there are limited numbers of NBEs and as R&D increases, more duplication will be observed), or a decreasing product yield.

Three alternative quasi-rent regimes are illustrated, one (Q1) with a weak IP regime, another (Q2) with a broader scope, and the third (Q3) with a broad

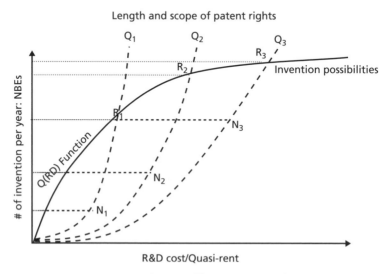

Figure 7.6. How stronger patent rights may affect the number of inventions
Source: Adapted from Scherer (2002).

and strong geographical scope. All three regimes are subject to diminishing marginal returns, as profit-seeking firms will try first the NBEs with the largest prospective payoffs. The Nordhaus' 'prospect' view of the world implies that firms choose projects without competition so as to maximize their net profits—i.e., the longest distance between the R&D cost function and the quasi-rent function. With a low payoff induced by weak IP rights (Q1), the profit maximizing equilibrium is at N1 NBEs. Strengthening the patent system (either in its scope or in its geographical reach) would lead to a sharp increase in inventions (to N2 or N3).

With the Barzel view of the world, there is a strong competition in R&D activities, and it therefore leads to much more R&D initially, up to the point where the quasi-rent function and R&D cost function intersect (R1), so that the net value is zero. More new products will also appear with a stronger patent system, but with clear diminishing return: going from Q2 to Q3 only leads to a small increase in NBEs, from R2 to R3.

In other words, this theoretical exercise suggests that the impact of a strengthening of the patent system depends on both the existing level of competition between innovators/researchers and the current strength of the patent system. A small level of competition would clearly justify a stronger IP policy. At the European level, one could therefore logically conclude that a stronger patent system (induced by the ratification of the London Protocol and the EPLA, for instance) would lead to more innovations if duplicative R&D across countries is reduced, which is a clear objective of the European Commission, and especially of DG Research.

7.5. **The Explosion of the EPO Workload**

The EPO has witnessed a radical surge in the size of its workload. Not only did the total number of patent applications increased drastically, but their size, as measured with the number of pages or number of claims per patent has also nearly doubled since the early 1980s. Table 7.6 illustrates the dramatic surge that took place over the past two decades in both the number and the size of patent filings.

The most striking increase took place in the number of patent applications at the EPO. From about 20,000 in the early 1980s, patent filings have jumped to 192,000 in 2005 and most probably more than 200,000 in 2006. Geographical and technological factors partly explain this evolution. The fastest growth has been observed for patents invented in Japan and the rest of Asia. The compound annual growth rate over the last 25 years was of more than 13 percent for Japanese applications, and more than 15 percent for the applications coming from the rest of the world (mainly the Asian region). This evolution will

Table 7.6. Evolution of EPO workload since 1980

	1980	1990	2000	2005
Total patent filings, incl. Euro-PCT filings	21 000	65 000	130 000	192 000*
Aver. Claims per patent	10	12	17	20*
Aver. Pages per patent	n.a.	16	27	30*
Total claims (000s)	210	780	2 210	3 840*
Total pages (000s)	n.a.	1 040	3 510	5 760*

Note: *The average patent number and size for the year 2005 are rough estimates.

Sources: EPO and Archontopoulos et al. (2006).

Box 7.3. The RIM case, part 4: patent flooding

RIM Woke up as consequence of this (and other) patent related trials, and now is flooding the USPTO and the EPO with many patent applications. In mid-April 2006 we performed simple patent search on the EPO and USPTO databases. The results presented here are illustrative of RIM's recent strategy, probably as a reaction to the NTP case.

Table 7.7. RIM's patenting behaviour in the US and Europe: number of patents

Date of	USPTO (filed)**	USPTO (issued)	EPO (filed)
1997	n.a.	3	0
1998	n.a.	6	0
1999	n.a.	5	0
2000	n.a.	9	0
2001	n.a.	9	0
2002	4	20	0
2003	9	20	4
2004	119*	17	56
2005	39*	27	242
2006	n.r.	27*	86*

Notes: *Indicates incomplete data. The search was performed at the end of April 2006. As patent applications are published 18 months after their priority filing, there is no information for the number of patent filings at the USPTO for the year 2006, and some 'missing' information for the year 2005. As RIM is based in Canada, most of its priority filings should be at the Canadian Patent Office, which explains the delay in the availability of recent information; **Patent filings at the USPTO were not published before 2001.

Since the early 2000s, RIM has drastically increased its number of patent filings, both in the US and in Europe, as illustrated in Table 7.7. The preliminary figures for the year 2006 suggest that a strong increase will occur in 2006 as well. So far, RIM has 388 patent applications at the EPO and 142 patents granted by the USPTO. This behaviour, which is more and more common nowadays, partly explains the sharp increase in patent applications worldwide. The *RIM v. NTP* case (see Box 4.3) and the *RIM v. InPro* case (see Box 7.3) seem to have induced RIM, which is a highly specialized company, to flood the market for technology with its patent applications, probably as a defensive reaction.

(Patent searches at USPTO (www.uspto.gov/) and EPO (http://ep.espacenet.com/))

probably continue at a similar pace, as China is becoming an important player in the field of patents. From 2000 to 2005 the domestic patent applications at the Chinese Patent Office (SIPO) have doubled, to reach more than 100,000 filings per year. Needless to say, this dynamism will translate into international applications.

On the technological side, biotechnologies and information and communication technologies were the most dynamic sectors in terms of patenting over the past ten years. For instance, the compound annual growth rate of applications arising from the information and communication technologies was of about 15 percent. Besides, a higher propensity to patent inventions has been observed for all technologies and in all countries, mainly driven by the strategic factors described in Chapter 4 (see Box 7.3).

This tenfold increase in the number of patents has been accompanied by a parallel evolution, although to a lower extent, of both the average number of pages and the average number of claims per filing, which have doubled between 1980 and 2005. The number of claims (pages) per patent was of about 12 (16 pages) in 1990, and is nowadays of 20 (30 pages). Nearly four million claims are applied nowadays at the office, with nearly six million pages to analyse. Over the past 25 years the workload of the office has been multiplied by 20.

This drastic evolution raises a crucial workload issue for the EPO and an important quality issue for the European patent system in general. According to Philipp (2006), a large proportion of the recent patents describe a small improvement upon the state of the art. This increased volume makes determining the actual state of the art almost impossible. One consequence is that a vicious cycle has taken place, inducing many 'doubtful' applications being filed, which increases the workload, and hence may reduce efficiency.

To explain the evolution in patent voluminosity, there surely is a generalized trend towards more complexity and more information associated with new technologies and cumulative inventions (e.g., the user manuals of home electronic devices are much larger and complex nowadays than 15 years ago). Archontopoulos et al. (2006) provide ample quantitative evidence on several factors that seem to influence the volume of patent applications.

One of these factor is related to the filing route (see Chapter 6) adopted by the applicant. The international PCT applications (PCT filings designating the EPO as International Search Authority) are composed of 25 claims on average, against 20 claims for Euro-PCT in the regional phase (PCT applications transferred to the EPO), and 15 claims for Euro-Direct applications (patents directly filed at the EPO without going through the international PCT procedure). PCT applications are much larger than the PCT that are effectively transferred to the EPO (nearly five additional claims) for the regional phase. The latter also have five more claims than the applications directly filed at the EPO. In other words, a difference of about ten claims can be observed for the patents that follow different filing routes. The three filing routes have increased to a similar extent between 2000 and 2004, with an average of two additional claims over the period. These differences in the size of PCT and non-PCT applications reflect the impact of different patent systems and their harmonization. The first filings at the USPTO generally have a much higher number of claims than the first filings at the EPO. As a PCT filing is an international route that later allows to transfer the application in several regional or national patent offices, firms tend to apply for one single patents that will be transferred internationally, instead of two patents for different jurisdictions. In this respect, the PCT filings seem to be more adapted to the US patent system than to the European one.

A second important driver of voluminosity is related to sectoral specificities: some sectors apparently need more claims in their patent applications. For instance, the biotechnology cluster is by far the technological area for which the number of claims and the number of pages are the largest. In 2005, the patent filings in the biotechnology cluster were composed of 35 claims on average. It is followed by computer related technologies (26 claims) and pure and applied organic chemistry (25 claims).

The geographical origin of applicants is a third source of heterogeneity in the size of patents. For instance, the patents with a priority filing in the United States have by far the largest number of pages and claims, followed by Denmark and Great Britain. The priority applications from Spain, Italy, and Germany have the smallest size.

In addition to these broad factors, van Zeebroeck et al. (2006) have identified other determinants of volume through an in-depth quantitative analysis of all patent applications at the EPO. The additional factors that have a significant impact on volume include the filing strategy of the firm (described in Chapters 4 and in 6), its experience (the number of patent filed over the previous years), the number of inventors underlying the invention, the practice that consists in filing first at the USPTO before extending the patent towards the EPO (for European applicants), and the level of complexity of the patents, as measured with the number of IPC classes (International Patent Classification).

In the European procedure some fees are related to the size of patent filed at the EPO. Claim-based fees are required at the application stage for the patents with more than ten claims, and page-based fees are required for the granted patents with more than 30 pages. The fees based on the size of patent might not be sufficiently discriminatory, as the recent evolution of patent filings shows no drop in their voluminosity. As of May 2006, the EPO has made it clear that it would revise its fees policy to proactively react to the steady increase in the size of patent documents. Similar actions have already successfully been taken by the USPTO in December 2004, where a new patent fee structure was put into practice, including significantly increased fees on claims, as well as on excessive number of pages of filed applications.

The workload of the EPO will probably continue to increase in the future, as the factors underlying the increasing number of patent applications might not be compensated by the factors reducing it, except if drastic changes are implemented. The three factors that may contribute to a sharp increase in the number of patent applications are (1) the improved integration of the European market for technology through the London Protocol or the European Patent Litigation Agreement, which may improve the attractiveness of the EPO; (2) the potentially sharp increase in patent filings originating from fast developing countries (e.g., India and China); and (3) the arrival of new actors, like universities and public research institutions, with an increased propensity to patent their inventions. The opposite forces are related to the rigor of the patent system and its cost. The cost being already relatively high, except maybe for claim-based fees, it seems that only a more rigorous selection process would contribute to reduce the propensity to apply for numerous patents.

7.6. **Summary**

- Over the past 20 years academic patenting in Europe has constantly increased, at a faster pace than patent applications by the business sector. This evolution is the result of three driving forces: the emergence of new technologies for which the frontier between basic and applied research is blurred; the adoption of Bayh–Dole act-like regulations in most European countries, which allows government funded research to be patented; and the financial support of regional governments targeted towards the technology transfer offices of universities.

- Concerns are raised about the potential negative effect of patenting on academic research (more applied research and a decrease in scientific productivity). The empirical evidence, however, underlines a positive and mutually reinforcing relationship between academic patents and the quality and quantity of scientific publications.

- A more important issue, which increasingly is being questioned, is related to the extent to which research activities relying partly on patented inventions fall outside the scope of patent protection and hence are not considered as infringing activities. It does seem to be the case in Europe. Not only is the EPC more favorable to research exemption, but researchers seem also to be unaware of the legal implications of using patented research tools, they simply disregard existing patents.

- Calculating the cost of patenting is far from being straightforward, as several components are not easy to quantify and depend on the applicant's strategy for filing a patent: the number of claims, the number of pages, the route, the quality of external services, the desired speed of the granting process, and the targeted geographical scope for protection are several factors that influence the cost of a patent. A survey-based approach and a simulation approach can be used to assess the average cost of a patent.

- A European patent that is renewed for ten years would cost between €23,000 and €56,000, depending on the number of countries targeted, against about €12,000 and €8000 for the US patent system and the Japanese patent system, respectively.

- The cost per claim per capita index, which is seven to ten times higher in Europe than in the US, shows that there is a clear negative relationship between the cost of patenting and the demand for patents, suggesting a strong patent applications elasticity with respect to fees. What makes a Europe-wide patent more expensive is mainly the translation costs (including intermediation with patent attorneys) and the validation fees in each national patent office that occur just after the grant of the patent by the EPO.

- Several indicators suggest, however, that the EPO is more selective in its granting process (longer examination time, smaller granting rate, more restrictions on subject matter, and an important rate of induced withdrawals). The higher rigour of the EPO granting process would justify higher examination costs but not the high complexity and cost burden induced by the currently fragmented market for intellectual property.

- Two main solutions are being investigated by European policy makers. The London Protocol would reduce the deadweight losses induced by expensive translations requirements, and the EPLA aims at the creation of a single and centralized European court. The strengthening of the European patent system would lead to more innovations if duplicative R&D within the EU is reduced.

- An explosion in the EPO workload has been observed over the past 20 years, characterized by both an increased number of patent filings and a constant increase in their size (number of pages or number of claims). The increased number and size of patents is lead by new technologies, emerging economies, a higher propensity to patent academic inventions, the adoption of various strategies by the business sector, and the international harmonization of patent systems.

- The workload of the EPO will probably continue to increase in the future, as the factors underlying the increasing number of patent applications might not be compensated by the factors reducing it, except if drastic changes are implemented. The improved integration of the European market for technology, the potentially sharp increase in patent filings originating from fast developing countries and the arrival of new actors (universities) with a higher propensity to patent their inventions, may all lead to a further increase in patent filings. The opposite forces are related to the rigor of the patent system and its cost. The cost being already relatively high it seems that only a more rigorous selection process may reduce the propensity to apply for patents.

8 The European Patent System at the Crossroad

Dominique Guellec and Bruno van Pottelsberghe

> Patents should draw a line between the things which are worth to the public the embarrassment of an exclusive patent, and those which are not. Patents are, after all, government-enforced monopolies and so there should be some 'embarassment' (and hesitation) in granting them.
>
> (Thomas Jefferson [1784] 2004)

This chapter reviews some of the major challenges currently faced by the European patent system and puts forward tentative responses. What is at stake is the ability of the patent system to continue fulfilling its mission in Europe (i.e., encouraging innovation and the diffusion of technology). While the mission has not changed, the context is different, hence the constraints faced by the system and the means through which it can operate. New technological fields are emerging (e.g., software and biotechnology); new actors (e.g., China and India); new market strategies (resulting in 'patent flooding'); new organizations of innovation (e.g., start-ups, cooperation between firms, and vertical disintegration); and the unfinished economic and political integration of Europe. The current patent system was not designed for operating in such a world, and confronting the corresponding challenges will require significant changes. Explicitly taking the economic dimension into account will help to find solutions, and possibly to find them faster.

8.1. The Twin Challenges of Quantity and Quality

The past 20 years have seen a surge in both the number and the size of patent filings. This explosion has induced a significant backlog issue for all patent offices in the world as they could not cope with such an inflated workload, creating internal tensions as the resources (examiners) were put under pressure to increase their productivity. In addition, the bulk of the surge has occurred in new technical areas, such as information technology, software, and biotechnology, in which there were initially fewer examiners and where patenting standards had to be adapted or even created—which obviously takes more resources. The economic forces which have generated and shaped this 'patent explosion' are here to last—globalization, new actors, competition, and the knowledge-based economy. Following the recent trends, the number of patent applications will not decline in the foreseeable future, and it will not even grow at a slower rate than the economy (which is around 3 percent worldwide), and with it the stress put on the patent system will be increased further. This cannot be neutral vis-à-vis quality, both on the side of applicants (quality of applications filed) and of patent offices (quality of search for prior art and of the substantive examination). An increase of applications which outpaces for so long the growth of economic variables such as R&D or GDP must generate and reflect a reduction in the average quality of these applications, except for a surge in the quality of R&D, which has, however, not been witnessed. For patent offices, managing an inflated workload can be done only by reducing the resources allocated to each application, mainly examiners' time, with the danger of substituting quantity for quality. In addition, a growing backlog also means increasing delays in the procedure: it takes ever longer for a patent application to be examined, which is a source of uncertainty both for applicants and for their competitors. Reducing the backlog would allow to reduce procedural delays. Concerns about the quality of patents are even raised by applicants themselves (e.g., Hagel 2004). What is more, a vicious causality seems to have occurred: as the patenting standards were being lowered, applicants felt encouraged to file ever lower quality inventions (as the probability of obtaining a patent for those was becoming greater), which in turn put more stress on patent offices, etc. When patenting standards, notably the inventive step, are lowered, it leads to inventors being obliged to file for more patents, as otherwise they incur the risk of small improvements on their own inventions being patented by others. Hence the only effective and economically 'safe' way of mastering the patent flood is to raise patenting standards: this would not only mitigate the problem faced by patent offices, but it would certainly help to improve competitive conditions on innovative markets.

From an economic perspective, a 'good patent system' should encourage inventions and should not hamper the diffusion of technology. Ideally, no patent should be granted to an invention that can be performed and publicized without a patent. Fulfilling this objective is a question of substantive patent law and case law and of procedural rules and practices. The system should be selective enough (rejecting 'bad applications'), but not too selective (not rejecting 'good applications' as well). In view of the current flood in applications, more emphasis should be put on the former aspect: selectivity should be reinforced. How can that be achieved?

• Applications benefit from a de facto presumption of validity, and the task of the patent office is to check it. As a consequence, it is much more difficult for the office to refuse an application than to grant it. A refusal requires detailed explanations by the examiner, which induces a longer processing time. Being under pressure to increase productivity, examiners are accountable for their time and they refuse applications only in a few cases—about 5 percent of applications at the EPO. This cost in refusal gives applicants an advantage in negotiating with examiners to have a patent (possibly with reduced breadth) granted. The role of examiners is, however, more important than the low refusal rate seems to show, as more than 50 percent of withdrawals (i.e., about one application in six) are induced by a communication from the examiner (either with the search report or during the substantive examination, see Lazaridis and van Pottelsberghe 2002). Refusals should be made easier for the examiner, by increasing the rewards they provide in accordance with the time and effort refusals usually require. That would reduce the current de facto bias for granting a patent. Further rewards could be added for each additional communication that takes place during the substantive examination process, keeping in mind, however, that each additional communication increases the granting (withdrawal, or refusal) delay by an additional year.

- The patenting procedure should be made less flexible and more rigorous, leaving a smaller margin for manoeuvre to applicants to change their applications during the process—in the form of amendments or divisionals. As applicants can enter into an endless game (a war of attrition) with examiners, until they are finally granted a (narrower) patent, implementing stricter rules would help discipline those who abuse the current rules. An interdiction of cascading divisionals, or restricting the number of these to one generation with a given time limit, would hamper manoeuvres aimed at deceiving examiners and competitors while gaining time. In addition, each divisional application should be examined before its publication to ensure that no additional subject matter is added to the earlier application (see Lang 2006).

- The quality of applications needs to be monitored, as applications of lower quality impose more work on the office, and further costs to society (hampering disclosure and competition while they are pending). Measures taken to this end could include both sticks and carrots: reducing the size of applications by applying claim-based fees as successfully implemented by the USPTO in December 2004 (the applicants pay more for a longer application); direct rejection of applications which lack clarity (in terms of language, structure, complexity, and citations); enforcing conditions for the clarity of applications which, when breached, would trigger a loss of priority for the applicants (i.e., they she cannot come back with a revised version while enjoying the protection awarded by the initial application); an application with clear claims (fulfilling the other criteria) could be granted rapidly, based on the preliminary opinion report (EESR or WOISA), and the applicant could have a rebate on the fees.

- Encouraging external contributions to the search for prior art: EPC Article 115 authorizes third parties to communicate relevant documents to the examiner in charge of a particular application. It is currently not frequently used, probably because third parties prefer to keep their information for an opposition procedure, where they can control its use, instead of submitting it to an examiner and losing control of it. This might be due to the fact that contributions to the prior art by third parties is not a public information, and is therefore not officially recognized. Contributions by external parties can be particularly useful in emerging technical areas—where much prior art is not available from standard sources, or difficult to access. In January 2006, the USPTO has joined forces with industry (notably IBM) and with the open source community (OSDL) for developing a software that would collect prior art on published patent applications from registered volunteers, who are automatically informed when a new application in their field of competence is published. The examiners then have to give recognition to the contributor, which guarantees that they have paid attention to the information. This same group, allied to Sourceforge.net (an open source group which operates a database of computer code), is also setting up a database of source code: a 'code repository' that will facilitate search for software related patent applications.

- Increasing the inventive step, e.g., by recognizing the higher skills of the 'person skilled in the art' or by giving more room to the examiner to invoke the 'general state of knowledge' instead of a specific document (sometimes difficult to find or not available) for rejecting applications. This was done in the software area in 2004 after the Hitachi decision taken by the Board of Appeal of EPO. Similar steps could be taken in other areas.

- Making the existing prior art more accessible to all users (individuals, SMEs, and universities). Large firms have a serious competitive advantage regarding the search and treatment of existing information regarding the prior art of the subject matter. Advanced search tools for patent and non-patent literature access and information retrieval would help small applicants in their search of prior art and in the drafting of their application. The search tools and databases currently available to examiners could be made public, with a free online access through the web for instance.

8.2. **The Policy Challenges**

The use of patents as a policy instrument for encouraging innovation and diffusion is in reality not straightforward in this era of globalization. Since the Paris convention of 1883, patent procedures should not discriminate between nationals and non-nationals: as about 50 percent of the applications received by the EPO are from non-European applicants, specific measures taken by the EPO will affect non-Europeans and Europeans equally. The patent system of a particular country or region might influence innovation patterns in that region, but also in other regions and conversely inventors in a given country are affected by the patent system of other countries. This is all the more true in emerging industries such as ICT or biotechnology, where markets are globalized: hence European companies can be at least as affected by a change in US patent law as by a change in European patent law.

Nevertheless, patent policies go beyond the granting procedure, and particular measures can be designed to reach certain objectives or certain types of inventors:

- SMEs and universities are confronted with various barriers when attempting to access the patent system. A major one is cost, to which small applicants, faced with liquidity constraints, are more sensitive than large ones. One solution is to reduce fees for small companies and/or universities, as in the US and Japan for instance, where a 50 percent discount is applied to 'small entities'. This policy would require a proper definition of SMEs, such as the one used by the European Commission (based on the number of employees and turnover).

- While encouraging universities to patent their inventions, it should also be ensured that this happens only when a patent is useful in stimulating downstream developments and broad applications of the technology, and that the patent does not constitute an obstacle to follow up inventions. This requires monitoring the potential negative impact of academic patents on the number and quality of academic publications, while maximizing the utilization of these patents through the creation of spin-offs, or through the licensing of inventions to firms. An 'exemption for research use' mechanism does exist in the patent law of most EU countries, based on the draft Community Patent regulation. This exemption allows third parties to make research on a patented invention' subject matter (i.e., to test or improve on the invention) without requesting authorization to the owner. It would be beneficial for research in Europe that the European patent law expands the research exemption for university inventions funded by government so that it includes research tools as well. Hence, research tools invented by universities through government funding would not be protected when they are used for doing research. This would reduce the private value of such inventions, but the motivation for government to fund them is not to maximize their private value. And, as nearly all difficulties in that field have come from university inventions (which have then been exclusively licensed or sold to a private company), that would solve the problem. This would be a form of government subsidy to the users of research tools invented by universities.

- Articulate the patent system with competition policy. Certain practices by patent holders are harmful to competition and society. The constitution of patent thickets, for instance, forecloses markets to new entrants. As such practices are at the interface of patent and competition law, a close cooperation between the bodies in charge of these two fields would be necessary. A more economic approach by patent bodies would help, as these problems are of economic nature. Competition authorities have already moved to this approach. It would create a common language to the two parties, with shared goals and shared conceptual framework that facilitate cooperation. On which grounds should such co-operation be established? (OECD 2004b) The control of the patent system should remain with patent offices, as competition authorities do not have the skills and resources needed to manage it. In addition, too much interference or overlap between the two bodies could increase legal uncertainty and cost, hence backfiring against innovation. The two groups of bodies could consult each other more systematically, exchange information (e.g., through drafting reports such as the FTC did on the US patent system in 2003), conduct joint or mutual training of their staff, etc. For instance, patent judges could be trained also in competition law, or some of them could be recruited with a background in competition law. The two groups could exchange more information, either of general interest, or concerning specific cases (like 'friends of the court' or 'Amicus curiae briefs': testimonies submitted by competition experts to patent courts or by patent experts to competition authorities). Competition authorities could also be encouraged to review granted patents, and file opposition when they deem particular patents harmful to competition while competitors themselves do not oppose (which could signal collusive behaviour).

8.3. **The Challenge of Integrating the European Patent System**

The European patent system is fragmented. It is made of a common grant procedure and an organization (the EPO) to implement it, plus a set of common rules that are applied in similar, but not necessarily identical ways in all countries. In Europe, patents are first of all national rights, submitted to national laws, enforced by national courts, and registered to national offices. That means differentiated rights across countries, a multiplication of filing costs, and the absence of a European policy in the field. The existence and success of the EPO manifests the willingness of Europeans to continue the process of integration. However, little progress have been made since the constitution of the EPO itself, except for an increase in the share of European patents as compared with national patents, and for an expansion in the membership of the EPO. Most attempts since the creation of the EPO to deepen European integration in the field of patents have failed. That includes most recently the Community Patent (CP) and the CII directive (computer implemented inventions) pushed forward by the European Commission. Why this repeated blockage?

The attempt to set up the CP failed in 2004 due to disagreement between governments over translation issues. Translation is a very sensitive area in some countries, notably France and Spain. It reflects real concerns regarding the dominance of a single language, English, in the field of patents. It reflects also the strength of particular interest groups. Some professional networks (notably translators and attorneys in certain countries) would lose from greater integration, as most of their revenue comes from foreign patentees who need to comply with local rules, including publication by the national patent office with translation requirements. These artificial barriers are clearly a toll to be paid by inventors to well connected groups. Another issue is the narrow and myopic views induced by the national interest of several governments. For instance, some countries, and particularly the ones which are less inventive but have a large market, have no immediate interest in a reform which would strengthen patents (see Scotchmer (2004) for the corresponding theory). Local customers would have to pay a higher price for patented goods (otherwise imitated by local producers), and local inventors would be submitted to a more intense foreign competition. A European system would protect more effectively the interests of the patent holder than most national systems do (stronger law, stronger courts). This explains the rejection of harmonization by a number of countries, which prefer to opt for a kind of 'patent protectionism'. This attitude is ill-founded. Customers might gain in price for certain goods, but this will be limited as the absence of protection also means that potential competitors will not supply the market, hence reducing competition and depriving customers of many new goods. The conservative attitude adopted by some European governments is to be contrasted with the proactive attitude of countries like China or India. These countries still produce relatively little innovation, they pay more in royalties to foreigners than they obtain, but they aim at becoming more innovative, and for that purpose are setting up a strong patent system, and have started filing internationally through the PCT process.

How can those with a vested interest be won over, making the interests of Europe, with all its member countries, prevail in the long run? Inscribing patent policy into the global dynamics of European integration—the Lisbon agenda— is the only way to show politicians and citizens what is at stake. Until now patent policy has been kept aside from the Lisbon process. This process is an economic one, while patent integration is seen as a legal process: that is precisely what has to be changed. The patent integration process must be given an economic motive as well. It is not just about completing treaties and harmonizing laws, it is about fostering innovation and access to technology in Europe. As the debate has been essentially legal until now, vested interests could more easily prevail, defending participants' 'rights' (legally founded) while ignoring the common good. In addition, legal systems are still strongly national, and it is difficult to find compromises between legal infrastructures that are so complex and different. Putting the patent system in to an economic dynamic would refocus the debate from legal details to the mission of the system, an area where different criteria apply and where different parts of the government are involved. There would be more likelihood that the European integration process would resume.

In what direction? What should an integrated European patent system look like? Initially, the EPC of 1973, instituting a 'European patent', was conceived as a transitory agreement until the implementation of the 'Community Patent' (CP) (the Luxemburg convention) was set in place. All attempts to date to implement the CP have failed, including the latest one which was in fact terminated in 2004. Not only did the CP fail, but also the European Commission could not push through its other harmonization projects related to patents. The software patents directive and the CP were both blocked by the European Parliament and Council, respectively. The biotechnology directive was voted in by Parliament in 1998, but it has not been implemented in a consistent way in all countries; France and Germany, in particular, have restricted the enforceable scope of patents on genetic material. In the absence of a European Court for patents, the EPC is being implemented in partially diverging ways in different countries, notably regarding the scope of patents and infringement.

EPO member states have tried to remedy this lack of integration at Community level by promoting EPC-based intergovernmental projects: the London protocol (which would reduce the translation requirements in signatory states) and the European Patent Litigation Agreement (EPLA) (which would set up an integrated judicial system for patents in Europe). These agreements would complement the EPC system, expand its scope, and make it more efficient. Such an approach has the merit of building on existing, working solutions. It will achieve its goal if it keeps close ties with the whole '*acquis communautaire*' (i.e., if the patent system develops in close association with other European institutions). A European patent court for instance could not be set up independently of the European Court of Justice (ECJ), not least because competition matters, which are sometimes closely related to patents, are treated there. More globally, the separate development of the patent system from the rest of the European institutions will have to be reviewed in the light of the missions assigned to the patent system. You cannot have patent matters at the heart of the European economic and political agenda when patent institutions still have a separate line of development and a large degree of independence.

8.4. **The Challenge of Reinforcing the Economic Dimension**

The patent system has to rely more heavily on economics. In doing that, it would follow the steps of competition law and policy, which started doing the same two decades ago in the US and more recently in Europe. The impact on prices and customers' choice (or welfare) has become the main criterion for assessing mergers and collaborative agreements between companies, instead of only looking at the level of market shares or the mechanical application of other pre-defined rules. In the words of Susan DeSanti of the Federal Trade Commission (2005):

The FTC recommended that patent practitioners similarly expand their consideration of economic learning and competition policy concerns in patent law decision making. The Supreme Court has made clear in several decisions that there is room for policy-oriented interpretation of the patent laws. Indeed, to find the proper balance between patent and competition law, the FTC stated, such policy-oriented interpretations are essential. Over the past twenty-five years, the incorporation of economic thinking into antitrust has provided significant insights that have substantially improved the development of antitrust law and competition policy. The Federal Circuit and the PTO may also benefit from much greater consideration and incorporation of economic insights in their decision making.

However introducing a greater focus on economics in the patent system is not straightforward, as one does not want to make patenting a discretionary or uncertain process. There is a need for clear rules and criteria which ensure legal certainty, without which the incentive effect of patents does not work. The system must be based on law, but the law must integrate economic concerns and its interpretation must take place in the light of economic considerations when those are relevant. In order to reach that stage, a preliminary step would be to add a preamble to a revised EPC, which would indicate the missions of the convention, in the light of which the convention should be interpreted. That revision could be complemented by a series of changes in the governance of the system aimed at ensuring that a broader set of concerns and interests are taken into account.

Inscribing the missions of the system in the EPC would fulfil several objectives. It would first serve as a compass in all adaptations of the system required by the emergence of new technologies, by changing strategies of applicants, or by the progress of European integration. It would provide guidance to judges when they take decisions over the scope of protection and patentability. It would be particularly useful when the system is confronted with new challenges that were not anticipated by those who drafted the EPC, and which require in fact judges to make the law. The system has already adapted in the past in reaction to external changes. But it took time, and this is more of a problem now than before as the pace of change has accelerated in all fields of society; shocks tend to be more frequent. A preamble laying down the missions of the system would allow quicker adaptation when new circumstances justify it.

Contrary to the EPC, the various EU directives and draft directives which have attempted to regulate patents in Europe have recitals. A prominent example is the TTBE (Technology transfer block exemptions; EC N. 772/2004) regulation, which extensively defines its scope and objectives in economic terms. Recital 1 comments, 'this is consistent with an economics-based approach which assesses the impact of agreements on the relevant market'. Instead of listing clauses forbidden under the Treaty of Rome, as did the 1994 regulation on technology transfers, the 2004 regulation defines categories of agreements exempted up to a certain level of market power. It would even prescribe different thresholds of market share depending on whether the licensing relationship is horizontal or vertical. In addition that regulation refers to the mission of the patent system, as its recital 14 reads: 'In order to protect incentives to innovate and the appropriate application of intellectual property rights, certain restrictions should be excluded from the block exemption'.

The proposed draft regulation of the Community Patent (CP) includes recitals which define the purpose of the CP as contributing to the completion of the single market, of which a single patent is a necessary component: that is true, but that does not define what the specific mission of a CP is, as compared with other components of the single market. The rest of the recitals in the CP draft regulation refer to the functioning of the system (sharing responsibility between EPO and national offices, etc.) The EC draft directive on the patentability of computer implemented inventions (CII, Doc. 7230/04 of the Council of the EU) also goes somewhat further:

The realisation of the internal market implies the elimination of restrictions to free circulation and of distortions in competition, while creating an environment which is favourable to innovation and investment. In this context the protection of inventions by means of patents is an essential element for the success of the internal market. Effective, transparent and harmonised protection of computer-implemented inventions throughout the Member States is essential in order to maintain and encourage investment in this field.

(Recital 1)

This is more specific than in the CP draft regulation but still quite short. The point in this recital is that 'effective, transparent and harmonised protection' is a necessary condition for encouraging invention: it does not state that encouraging inventions is the goal of the system. The patent system is still viewed as an independent entity, which does not need to be assessed with regards to its impact on inventions, but which should be adjusted *internally*, according to its own logics, and would result fortunately in encouraging inventions. This is different from saying that the patent system should be evaluated on the basis of its effect on inventions, and if needed, reformed accordingly. In addition, it is odd (and legally strange), to have the general missions of a system defined only in a regulation concerning one particular type of application: does it mean that the missions of the patent system, say for chemistry, would not be the same as for CII? The same question applies to the directive on the protection of biotechnological inventions of 1998 (98/44/EC), which spells out clearly and extensively the role of patents for encouraging inventions and industrial development. The preamble to this directive is the best precedent for a preamble to the EPC, of which we propose a draft in Box 8.1.

Box 8.1. Draft proposal for recitals to a revised EPC

This series of recitals would be placed as a preamble to the EPC.

1. The European patent system as designed in this convention should meet the two requirements of encouraging inventions and the diffusion of knowledge in the European internal market, while providing adequate legal security of undertakings. The pursuit of these objectives should take account of the need to preserving competition and respecting ethical values of European societies. Institutions of the European patent system will ensure adequate inclusion of economic and other social objectives in their decision making process.

2. A patent shall confer to its owner the rights to prevent third parties not having the owner's consent from making, using, offering for sale, selling, or importing for these purposes the patented product (or product obtained from a patented process).

3. European patents are granted by the EPO with the view to provide incentives to investment in innovation and exploitation of inventions, and to the diffusion of technology. The conditions of protection provided by patents, such as the scope and the inventive step, should be sufficient to ensure proper incentives to inventors and investors, including compensation for the risk taken.

4. Patents can sometimes reduce competition as they restrict the rights of competitors to the benefit of the inventor. Standards for granting patents should be set so as to minimize such effects, allowing competitors to enter markets while preserving incentives to inventors. Patents can also enhance competition, by encouraging new entrants, with new technology, to compete with incumbents which control markets due to other factors such as reputation, etc.

5. Patents must enhance the diffusion of knowledge. Inventions must be described in patent documents in a way which makes them accessible to third parties who want to use them for further inventions. The scope and other aspects of protection granted should not deter derived inventions.

6. Patents are a source of benefits and costs to their holder and to third parties. They are justified on the ground that their benefits to society as a whole more than outweigh their cost. The governance of the European patent system must reflect that concern. In particular, all significant stakeholders present in society should be involved in policy decision making, including the business sector, public research organizations, government, customers, and citizens.

7. Patents are used by their holders for a variety of purposes, including licensing (which serves diffusion) and raising capital (which serves exploitation). Decisions on European patents taken on the basis of the convention will integrate these aspects when it is relevant.

8. Patent applicants and their representatives have the right to be heard by the EPO. However, such a right should not be used in a way which allows the abuse of the rules of the system and circumventing the aims of this convention at the expense of other parties. The EPO should take proportionate measures so as to deter such strategies.

9. The patent system is part of the establishment of an internal market characterized by the abolition of obstacles to the free movement of goods and of distortions to competition. It serves the establishment of Europe-wide policies aimed at encouraging innovation and the diffusion of knowledge. The governance of the European patent system should ensure consistency of patent processes with European innovation policy, European competition policy and the Commission bodies in charge of implementing them.

10. The integration of economic concerns in judicial decisions taken on the basis of this convention should be compatible with the legal security needed by inventors and other parties.

Would this mean that a further criterion, 'usefulness to the economy', should be added to existing statutory criteria for granting patents? No. Economics does not in general provide precise enough guidance to allow such a move while not introducing much legal uncertainty, and existing statutory criteria provide a proxy which is sufficient in most cases for ensuring that granted patents will not have a detrimental effect on the economy. It means that in fields and cases of particular relevance, the existing criteria should be implemented with economic concerns in mind. Issues such as the secondary use of drugs and the patentability of CII should be settled by political authorities and the courts on the basis of articulated economic reasoning and evidence. The current situation would then be changed so that economic considerations would be strengthened (notably at the expense of exegesis of existing texts), and economic arguments grounded (whereas today they are already often invoked, but rarely substantiated). Regarding individual cases, economic concerns should be invoked, when necessary, to calibrate the implementation of statutory criteria. In the same way as the scope of a patent would be reduced in view of, for example, the balance between the suffering imposed on animals and the eventual benefits to humanity (EPO Board of Appeal decision T 0315/03 of 6 July 2004, on the oncomouse patent case), consideration could be taken, when appropriate, of a balance between the positive and negative effects on the economy, the incentive to the inventor, and the reduction of access and competition. This balance is not easy to strike, but when cases reach the Board of Appeal the invention generally has been marketed for several years already, and so the experience of what happened since the patent application was filed could be used by judges. This would not require more speculation than estimating the suffering of animals.

In order to substantiate the integration of economic concerns in the patent system the following measures would help:

1. Broadening the set of interests involved in patent policies: there are already channels for direct stakeholders in the system to be heard—such as the SACEPO, an advisory body to the EPO made mainly of professionals

(applicants and representatives). It is noticeable, however, that consumers' associations are not very active on IPR issues: very few initiatives can be mentioned (such as the Consumer Project on Technology, led by consumers' activist Jamie Love in the US). This is regrettable as consumers are those who pay for the system (through the markup on the price of patented goods), and they are supposed to be the ones who benefit from it also (i.e., with new products broadening choice and new processes lowering cost). Hence, it would be politically fair to invite selected consumers' associations to be part of the consultative bodies involved in the governance of the patent system, giving them equal weight as direct users of the system (notably attorneys).

2. Carry out further empirical studies investigating the economic impact of various aspects of the law and of particular legal decisions. Such studies would provide policy makers (e.g., the European Parliament) and judges with a basis on which they could ground some of their decisions. They would also inform the public and stakeholders, allowing public debates of high standard to be held, based on evidence. Regular evaluation of the patent system should be conducted, making as much use as possible of quantitative indicators (e.g., impact on innovation, diffusion, competition, and prices, etc.) A similar clause was put in the biotechnology directive (Article 16) and the CII draft-directive, according to which the voted law should be evaluated after a period of five years: this approach should simply be extended to other aspects of patent law and policy. Such measures would contribute to weaken the grasp of vested interests on the patent system, and to shape it so that it could fulfil its policy mission in the twenty-first century.

■ REFERENCES

AARP, Public Patent Foundation, and Consumers Union (2005). *Brief to the Supreme Court of the US as Amici Curiae*. Available at: http://www.pubpat.org/PUBPAT_Ind_Ink_SCt_Brief.pdf

Angell, M. (2004). 'The Truth about Drug Companies'. *The New York Review of Books*, 51(12).

Archontopoulos, E., Guellec, D., Stevnsborg, N., van Pottelsberghe de la Potterie, B., and van Zeebroeck, N. (2006). 'When Small is Beautiful: Measuring the Evolution and Consequences of the Voluminosity of Patent Applications at the EPO'. EPO and Working Papers CEB 06-019. RS, Université Libre de Bruxelles, Solvay Business School, Centre Emile Bernheim (CEB).

Arora, A., Gambardella, A., and Fosfuri, A. (2001). *Markets for Technology*. Cambridge, MA: MIT Press.

—— Ceccagnoli, M. and Cohen, W.C. (2003). 'R&D and the patent premium'. NBER Working Paper, 9431.

Arrow, K.J. (1962). 'Economic Welfare and the Allocation of Resources for Invention'. Universities: National Bureau of Economic Research Conference. Princeton, NJ: Princeton University Press.

Arundel, A. (2001). 'The Relative Effectiveness of Patents and Secrecy for Appropriation'. Research Policy, 30: 611–24.

—— and Kabla, I. (1998a). 'What Percentage of Innovations are Patented? Empirical Estimates for European Firms'. *Research Policy*, 27: 127–41.

—— and —— (1998b). 'Patenting Strategies of European Firms: An Analysis of Survey Data', in S. Allegrezza and H. Serbat (eds.) *The Econometrics of Patents*. Paris: Applied Econometrics Association Series.

Athreye, S.S. and Cantwell, J. A. (2005). 'Creating Competition? Globalisation and the Emergence of New Technology Producers'. *Open University Economics Discussion Paper*, 52. Available at: http://ssrn.com

Azoulay, P., Ding, W., and Stuart, T. (2005). *'The Impact of Academic Patenting on the Rate, Quality and Direction of (Public) Research Output*. New York: Columbia University.

Barzel, Y. (1968). 'Optimal Timing of Innovations'. *Review of Economics and Statistics*, 50(3): 348–55.

Beier, R. (undated). *Strategies to Defend a European Patent in the Opposition Procedure before the European Patent Office*. Available at: http://www.ip-firm.de/aufsatz1.pdf

Bekkers, R., Duysters, G., and Verspagen, B. (2002). 'IPR, Strategic Technology Agreements and Market Structure: The Case of GSM'. *Research Policy*, 31: 1141–61.

Belfanti, M. (2004). 'The Complex Role of Patents in Creating Technological Competencies'. Working Paper presented to the 4th EPIP Conference, Paris.

Beltran, A., Chauveau, S., and Galvez-Behar, G. (2001). *Des brevets et des marques: Une histoire de la Propriété Industrielle*. Paris: Fayard.

Bessen, J. and Maskin, E. (2000). 'Sequential Innovation, Patents and Imitation'. Working Paper. Cambridge, MA: Department of Economics, MIT.

Black, F. and Scholes, M. (1973). 'The Pricing of Options and Corporate Liabilities'. *Journal of Political Economy*, 81(3): 637–54.

Blind, K. (2004). '*The Economics of Standards: Theory, Evidence Policy*. Cheltenham: Edward Elgar.

Bloom, N. and van Reenen, J. (2002). 'Patents, Real Options and Firm Performance'. *Economic Journal*, 112(478): 97–116.

Bosworth, D.L. and Pitkethly, R. (forthcoming). 'IP Management, Technology and Business Strategy at Group Lotus PLC'. *International Journal of Technology Management*, 22.

Braendly, P. (Munich Diplomatic Conference for the Setting up of a European System for the Grant of Patents) (1973). *Report on the Decisions and Discussions of the Main Committee I*, IIC(4): 402–18.

Braunstein, P. (1984). 'A l'origine des privilèges d'invention'. Mimeo, EHESS, Paris.

Breschi, S., Lissoni, F., and Montobbio, F. (2006). 'From Publishing to Patenting: How to Become an Academic Inventor'. Working Paper. Cespri: University of Bocconi.

——— ——— ——— (forthcoming). 'The Scientific Productivity of Academic Inventors: New Evidence from Italian Data'. *Economics of Innovation and New Technology*.

Brooks, E.J. and Ware, C.R. (2006). 'Claim Strategies in View of the New Fee Structure'. Working Paper. Minneapolis: Brooks & Cameron, PLLC.

Brouwer, E. and Kleinknecht, A. (1999). 'Innovative Output and a Firm Propensity to Patent: An Exploration of CIS Micro Data'. *Research Policy*, 28(6): 615–24.

Burk, D. and Lemley, M. (2003). 'Policy Leverages in Patent Law'. Research Paper, 135. New York: University College Berkeley.

Cassier, M. and Correa, M. (2005). 'Patents, Innovation and Public Health: Brazilian Public Sector laboratories Experience in Copying AIDS Drugs'. Presentation at the EPIP Conference of Santiago. Available at: http://www.usc.es/epip/documentos/Cassier%27s%20paper.pdf

Cassiman, B., Pérez-Castrillo, D. and Veugelers, R. (2001). 'Endogenizing Know-how Flows through the Nature of R&D Investments'. *International Journal of Industrial Organization*, 20(6): 775–99.

Chiao, B., Lerner, J., and Tirole, J. (2005). 'The Rules of Standard Setting Organizations: An Empirical Analysis'. NOM Working Paper No. 05–05. Cambridge, MA: Harvard University Press.

CJA Consultants (2003). 'Patent Litigation Insurance: A Study for the European Commission on Possible Insurance Schemes Against Patent Litigation Risks'. *Final Report for the European Commission*, January. Available at: http://europa.eu.int/comm/internal_market/indprop/docs/patent/studies/litigation_en.pdf

Clarkson, G. (2004). 'Objective Identification of Patent Thickets: A Network Analytic Approach'. PhD dissertation, Harvard Business School, Cambridge, MA, p.120.

Cohen, W.M. and Levin, R.C. (1989). 'Empirical Studies of Innovation and Market Structure', in R. Schmalensee and D. Willig (eds.) *Handbook of Industrial Organization*. London: North-Holland.

—— Nelson, R.R., and Walsh, J.P. (2000). 'Protecting their Intellectual Assets: Appropriability Conditions and Why US Manufacturing Firms Patent (or Not)'. Working paper, 7552. Cambridge, MA: NBER.

Cornelli, F. and Schankerman, M. (1999). 'Patent Rewards and R&D Incentives'. *Rand Journal of Economics*, 30(2): 197–213.

Crampes, C., Encaoua, D., and Hollander, A. (2005). 'Competition and Intellectual property in the EU'. Working papers 332. Toulouse: IDEI.

Crépon, B., Duguet, E., and Kabla, I. (1996). 'Schumpeterian Conjectures: A Moderate Support from Various Innovation Measures', in A. Kleinknecht (ed.) *Determinants of Innovation: The Message from New Indicators*. New York: Palgrave.

Crépon, B., Duguet, D., and Mairesse, J. (1998). 'Research, Innovation, and Productivity: An Econometric Analysis at the Firm Level'. *Economics of Innovation and New Technologies*, 7(2): 115–58.

Dack, S.C. and Cohen, B. (2001). *Complex Applications: A Return to First Principles*, IIC(32): 5.

Dasgupta, P. and David, P.A. (1994). 'Towards a New Economics of Science'. *Research Policy*, 23: 487–521.

Daumas, M. (1964). 'A History for Technology and Inventions: Progress Through the Ages', vol. 3. New York: Crown Publisher.

David, P.A. (1993). 'Intellectual Property Institutions and the Panda's Thumb: Patents, Copyrights, and Trade Secrets in Economic Theory and History', in M. Wallerstein, M. Mogee, and R. Schoen (eds.) *Global Dimensions of Intellectual Property Rights in Science and Technology*. Washington, DC: National Academy Press.

Demsetz, H. (1968). 'Toward a Theory of Property Rights', *American Economic Review*, 347–59.

Dent, C., Jensen, P., Waller, S., and Webster, B. (2006). 'Research Use of Patented Knowledge: A Review'. OECD/DSTI, DOC82006(2): 52.

Duffy, J.F. (2004). 'Rethinking the Prospect Theory: A Neo-Demsetzian View'. *University of Chicago Law Review*, 71: 439.

Duguet, E. and Kabla, I. (1998). 'Appropriation Strategy and the Motivations to Use the Patent System: An Econometric Analysis at the Firm Level in French Manufacturing'. *Annales d'Economie et de Statistique*, 49(50): 289–327.

—— and Lelarge, C. (2005). Les brevets incitent-ils les entreprises à innover? *Economie et Statistique*, 380: 35–60.

Edwards, D. (2002). *Patent Backed Securitization: Blueprint for a New Asset Class*. Available at: http://www.securitization.net/pdf/gerling_new_0302.pdf

Encaoua, D. and Hollander, A. (2004). 'Price Discrimination, Competition and Quality Selection. *Cahiers de la MSE*. Paris: Université Paris I.

—— and Ulph, D. (2000). 'Catching-up or Leapfrogging? The Effects of Competition on Innovation and Growth'. *Cahiers de la MSE*. Paris: Université Paris I.

—— Martinez, C. and Guellec, D. (forthcoming) 'Patent Systems for Encouraging Innovation: Lessons from Economic Analysis'. *Research Policy*.

EPO (European Patent Office) (2001). *Case Law of the Board of Appeal of the EPO*. Munich: EPO.

EPO (European Patent Office) (2003). *Annual Report 2003*. Munich: EPO.

EPO (European Patent Office) (2004). *Decision T 0258/03 of the Board of Appeal* ('Hitachi decision'). Munich: EPO.

EPO (European Patent Office) (2005a). *Guidelines for examination in the European Patent Office*. Munich: EPO.

EPO (European Patent Office) (2005b). *Patents in Europe*. A supplement to *IAM Magazine*. Munich: EPO.

EPO (European Patent Office) (2005c). *Applicant Panel Survey 2004*. Internal report. Munich: EPO.

Falvey, R., Foster, N., and Greenaway, D. (2004). 'IPR and Economic Growth'. Research paper 2004/12, Unversity of Nottingham, Nottingham.

Farell, J. and Gallini, N. (1988). 'Second Sourcing as a Means of Commitment: Monopoly Incentives to Attract Competition'. *Quarterly Journal of Economics*, 108: 673–94.

Federal Trade Commission (2003). *To Promote Innovation: The Proper Balance between Competition and patent law and Policy*. Washington, DC: FTC.

Flynn, W. J. (2006). *Patents since the Renaissance*. Available at: Booklocker.com

François, J-P. and Lehoucq, T. (1998). 'Les entreprises face à la propriété intellectuelle'. *Le 4 Pages*, 86, SESSI. Paris: Ministère de l'Industrie.

Franzozi, M. (2003). 'Markush Claims in Europe'. *E.I.P.R. Comments*, 4: 200–3.

Fröhling, W. (2005). 'Practical Experiences Regarding the Evaluation of Medium-sized Patent Portfolios'. Paper presented at the International Conference on IP as an Economic Asset: Key Issues in Exploitation and Valuation, Berlin.

Gallini, N. (1992). 'Patent Policy and Costly Innovation'. *Rand Journal of Economics*, 21: 106–12.

—— and Scotchmer, S. (2002). 'Intellectual Property: When is it the Best Incentive System?' in A. Jaffe, J. Lerner, and S. Stern (eds.) *Innovation Policy and the Economy*. Cambridge, MA: MIT Press.

Galvez-Béhar, G-D. (2004). 'Pour la fortune et pour la gloire: inventeurs, propriété industrielle et organisation de l'invention, France 1866–1922'. PhD thesis, Université de Lille III.

Gambardella, A. (2005). 'Assessing the "Market for Technology" in Europe'. A report to the European Patent Office (EPO). Munich: EPO.

Garrisson, C. (2004). 'Intellectual Property Rights and Vaccines in Developing countries'. Background paper for WHO workshop, Geneva, 19–20 April.

Gilbert, R. and Shapiro, C. (1990). 'Optimal Patent Length and Breadth'. *Rand Journal of Economics*, 21: 106–12.

Ginarte, J.C. and Park, W.G. (1997). 'Determinants of Patent Rights: A Cross-National Study'. *Research Policy*, 26(3): 283–301.

Glazier, S.C. (2000). *Patent Strategies for Business* (3rd edn.) Washington, DC: LBI Institute, p.420.

Gould, D.M. and Gruben, W.C. (1996). 'The Role of Intellectual Property Rights in Economic Growth', *Journal of Economic Development*, 48: 323–50.

Graham, S., Hall, B.H., Harhoff, D., and Mowery, D.C. (2003). 'Post-Issue Patent Quality Control: A Comparative Study of US Patent Re-Examinations and European Patent Oppositions', in W. M. Cohen and S. A. Merrill (eds.) *Patents in the Knowledge-Based Economy*. Washington, DC: The National Academies Press, pp.74–119.

Grindley, P. (1995). *Standards Strategy and Policy: Cases and Stories*. New York: Oxford University Press.

Guellec, D. and Martinez, C. (forthcoming). 'Overview of Recent Trends in Patent Regimes in United States, Japan and Europe'. Working Paper. Paris: OECD.

—— and van Pottelsberghe de la Potterie, B. (2000). 'Applications, Grants and the Value of Patents'. *Economic Letters*, 69(1): 109–14.

—— —— (2003). 'The Impact of Public R&D Expenditure on Business R&D'. *Economics of Innovation and New Technologies*, 12(3): 225–44.

Hagel, F. (2004) 'La politique pro-déposant de l'OEB a des conséquences négatives'. *EPI Information*, 1: 29–31.

Hall, B.H. and Harhoff, D. (2004). 'Post Grant Review Systems at the U.S. Patent Office—Design Parameters and Expected Impact'. *Berkeley Law Technology Journal*, 19: 989–1016.

—— and Ziedonis, R.H. (2001). 'The Patent Paradox Revisited: An Empirical Study of Patenting in the U.S. Semiconductor Industry, 1979–95'. *Rand Journal of Economics*, 32(1): 101–28.

Hall, B., Jaffe, A., and Trajtenberg, M. (2001). 'The NBER Patent Citations Data File: Lessons, Insights And Methodological Tools'. NBER Working Paper, 8498. Available at: http://www.nber.org/papers/w8498

—— —— —— (2005). 'Market Value and Patent Citation'. *Rand Journal of Economics*, Spring, 36(1): 16–38.

Hammer, T. (2005). Proceedings of the Conference on the Quality in the European Patent System, EPO, the Hague, 21–22 November.

Hansen, S., Brewster, A., and Asher, J. (2005). *Intellectual Property in the AAAS Scientific Community*. Washington, DC: American Association for the Advancement of Science.

Harhoff, D. (2004). 'Patents in Europe and the European Patent System'. Paper presented at the European Parliament, Brussels, 10 November. Available at: http://www.ffii.org/~jmaebe/conf0411/wed/Dietmar%20Harhoff.pdf

—— (2005). 'Incidence, Duration and Outcomes of Opposition and Appeal Cases at the European Patent Office'. Discussion paper. London: Center for Economic Policy Research.

—— (2006a). 'The Battle for Patent Rights', in C. Peeters and B. van Pottelsberghe de la Potterie (eds.) *Economic and Management Perspectives on Intellectual Property Rights*. London: Palgrave MacMillan.

—— and Hall, B.H. (2004). 'Intellectual Property Strategy in the Global Cosmetics Industry'. Unpublished discussion paper, Munich School of Management, Munich.

—— and Reitzig, M. (2004). 'Determinants of Opposition against EPO Patent Grants: The Case of Biotechnology and Pharmaceuticals'. *International Journal of Industrial Organization*, 22: 443–80.

—— and Scherer, F.M. (2000). 'Technology Policy for a World of Skew-distributed Outcome'. *Research policy, 29:* 559–66.

—— and Wagner, S. (2004). 'Modeling the Duration of Patent Examination at the European Patent Office'. Unpublished discussion paper, Munich School of Management, Munich.

—— Scherer, F.M., and Vopel, K. (2002). 'Exploring the Tail of the Patent Value Distribution', in O. Granstrand (ed.), *Economics, Law and Intellectual Property: Seeking Strategies for Research and teaching in a Developing Field*. London: Kluwer Academic Publisher, p.279–309.

Hart, O. and Moore, J. (1990). 'Property Rights and the Nature of the Firm'. *Journal of Political Economy*, 98(6): 1119–58.

Hartel, H. and Kolb, A. (eds.) (1987). *IDC Chronik 1967–1987*. Sulzbach: IDC.

Haxel, C. (2002). 'Patent information at Henkel: From Documentation and Information to Collaborative Information Commerce'. *World Patent Information*, 24: 25–30.

Hayek, F. (1945). 'The Use of Knowledge in Society'. *American Economic Review*, 35(4): 519–30.

Heller, M.A. and Eisenberg, R.S. (1998). 'The Tragedy of the Anticommons'. *Science*, 280: 698–701.

Henderson, R., Jaffe, A.B., and Trajtenberg, M. (1995). 'Numbers Up, Quality Down? Trends in University Patenting 1965–1992'. Draft manuscript. Stanford: CEPR.

—— Jaffe, A.B., and Trajtenberg, M. (1998a). 'University Patenting Amid Changing Incentives for Commercialization', in G. Barba Navaretti, P. Dasgupta, K.G. Mäler, and D. Siniscalco (eds.) *Creation and Transfer of Knowledge*. New York: Springer.

—— —— —— (1998b). Universities as a Source of Commercial Technology: A Detailed Analysis of University Patenting, 1965–1988'. *Review of Economics and Statistics*, 80: 119–27.

Herrera, H. and Schroth, E. (2000). 'Profitable Innovation Without Patent Protection: The Case of Credit Derivatives'. Working Paper. Berkeley: Berkeley Center, Stern NYU.

Hilaire-Perez, L. (2000). '*L'invention technique au Siècle des Lumières*'. Paris: Albin Michel.

—— (2002). 'Diderot's Views on Artists' and Inventors' Rights: Invention, Imitation and Reputation'. *British Journal for the History of Science*, 35: 129–50.

Hillery, J. (2004). 'Securitization of Intellectual Property: Recent Trends from the United States'. Working Paper. Washington, DC: CORE. Available at: http://www.iip.or.jp/summary/pdf/WCORE2004s.pdf

House of Lords (2004). 'Opinion in the Cause Kirin-Amgen Inc and Others *[Appellants] v. Hoechst Marion Roussel*'. London: House of Lords.

Hunt, R.M. (1999). 'Nonobviousness and the Incentive to Innovate: An Economic Analysis of Intellectual Property Reform'. Working Paper 99–3. Philadelphia: *Federal Reserve Bank of Philadelphia*.

—— (2004). 'Patentability, Industry Structure, and Innovation'. *Journal of Industrial Economics*, 52: 401–25.

Husinger, K. (2004). 'Is Silence Golden? Patents versus Secrecy at the Firm Level'. Discussion Paper, 04: 78. Mannheim: ZEW.

IPR Help Desk (2004). 'Some Basic Issues Surrounding Improvements Made to Patented Inventions and to Dependent Patents.' Brussels: European Commission.

—— (2005). 'Patenting and the Research Exemption'. Brussels: European Commission.

Israelsen, N.A., Perdue, D., and Rickenbrode, J. (2002). 'Festo and Biotechnology Patent Prosecution: Consequences for the Practical Patent Practicioner'. *Biotechnology Law Report*, 21(6): 540–50.

Jaffe, A.B. (2000). 'The U.S. patent system in transition: policy innovation and the innovation process'. *Research Policy*, 29: 531–57.

Jaffe, A. and Lerner, J. (2004). *Innovation and its Discontents: How Our Broken Patent System is Endangering Innovation and Progress, and What to do About It*. Princeton: Princeton University Press, p.236.

—— Trajtenberg, M., and Henderson, R. (1993). 'Geographic Localization of Knowledge Spillovers as evidenced by patent citations'. Working Papers, 108/3. Cambridge, MA: NBER.

Jefferson, T. ([1784]2004). 'Monopolies of the Mind', *Economist*, 11 November.

Kaiser, J. (2005). 'Up Close: Finding an Alternative to the European Opposition System'. *Patent World*, 176: 21–4.

Kang, S.J. and Seo, H.J. (2006a). 'Do Stronger Intellectual Property Rights Induce More Patents?' in C. Peeters and B. van Pottelsberghe de la Potterie (eds.) *Economic and Management Perspectives on Intellectual Property Rights*. London: Palgrave MacMillan, p.129–45.

Kanwar, S. and Evenson, R.E. (2001). 'Does Intellectual Property Protection Spur Technical Change?' Discussion Paper N. 831. New Haven, CT: Yale University.

Keyworth, B. (2004). 'Better Patents Using Technical Writing Techniques'. *Cafezine*.

Khan, M. (2001). 'Investment in Knowledge'. *STI Review N. 27*. Paris: OECD, pp.19–47.

Khan, B.Z. (2006). 'An Economic History of Patent Institutions'. *EH.Net Encyclopedia*. Available at: http://eh.net/encyclopedia/article/khan.patents

Kingston, W. (2001). 'Innovation Needs Patent Reform'. *Research Policy*, 30: 403–23.

Kitch, E. (1977). 'The Nature and Function of the Patent System'. *Journal of Law & Economics*, 20: 265.

Klemperer, P. (1990). 'How Broad should the Scope of Patent Protection Be?' *Rand Journal of Economics*, 21: 113–30.

Kortum, S. and Lerner, J. (1998). 'Stronger Protection or Technological Revolution: What Is Behind the Recent Surge in Patenting?' *Carnegie-Rochester Series on Public Policy*, 48: 247–304.

Kothari, V. (2003). *Securitisation: The Financial Instrument of the New Millennium.* Calcutta: Academy of Financial Services.

Kraft, K. (1989). 'Market Structure, Firm Characteristics and Innovative Activity'. *Journal of Industrial Economics*, 37(3): 329–36.

Landes, W.M. and Posner, R.A. (2003). '*The Economic Structure of Intellectual Property Law*'. Cambridge, MA: Harvard University Press.

Lang, J. (2006). 'Divisional applications: flexibility versus certainty'. *Patent World*, March, 180: 15–18.

Lanjouw, J. and Schankerman, M. (1997). 'Stylised Facts of Patent Litigation: Value, Scope and Ownership'. Working Paper, 6297. Cambridge, MA: NBER.

—— —— (1999). 'The quality of ideas: measuring innovation with multiple indicators'. Working Paper, 7345. Cambridge, MA: NBER.

—— —— (2004). 'Protecting Intellectual Property Rights: Are Small Firms Handicapped?'. *Journal of Law and Economics*, 47: 45–74.

—— Pakes, A., and Putnam, J. (1998). 'How to Count Patents and Value Intellectual Property: The Uses of Patent Renewal and Application Data'. *Journal of Industrial Economics*, 46(4): 405–32.

Laub, C. (2006). 'Software Patenting: Legal Standards in Europe and the US in view of Strategic Limitations of the IP Systems'. *Journal of World Intellectual Property*, 9(3): 344–72.

Layton, R. and Bloch, P. (2004). *IP Donations: A Policy Review.* Washington, DC: International Intellectual Property Institute. Available at: http://www.iipi.org/reports/IP_Donations_Policy_Review.pdf

Lazaridis, G. and van Pottelsberghe de la Potterie, B. (2006). 'The Rigour of EPO's Patentability Criteria: An Insight into the "Induced Withdrawals"'. EPO and CEB Working Paper CEB 06-018. RS. Brussels: Université Libre de Bruxelles, Solvay Business School, Centre Emile Bernheim (CEB).

Lelarge, C. (2002). 'L'utilisation du brevet dans le processus d'innovation des entreprises européennes'. Rapport de stage ENSAE. Mimeo, ENSAE, Paris.

Lemley, M. (2001). 'Rational Ignorance at the Patent Office'. Working Papers 2000(2): 40. Berkeley: The Berkeley Law & Economics.

—— (2002). 'Intellectual Property Rights and Standard Setting Organizations'. *California Law Review*, 90: 1889.

—— (2004). 'Property, Intellectual Property, and Free Riding'. *Stanford Law School Working Paper*, 291.

Lerner, J. (1994). 'The Importance of Patent Scope: An Empirical Analysis'. *Rand Journal of Economics*, 25(2): 319–33.

—— (2002). 'Patent Protection and Innovation over 150 Years'. Working Papers, 8977. Cambridge, MA: NBER.

Lev, B. (2001). *Intangibles: Managemenmt, Measurement and Reporting.* Washington, DC: Brookings Institution Press.

Levin, R.C., Klerovick, A.K., Nelson, R.R., and Winter, S.G. (1987). 'Appropriating the Returns from Industrial Research and Development'. *Brookings Papers of Economic Activities*, 3: 783–831.

Licht, G. and Zolz, K. (1998). 'Patents and R&D: An Econometric Investigation Using Applications for German, European and US patents by German Companies'. *Annales d'economie et de statistique*, 49(50): 329–60.

Lipfert, S. and von Scheffer, G. (2006). 'Europe's First Patent Value Fund'. *Intellectual Asset Management*, pp.15–18.

Locke, J. (1690). *Second Treatise of Civil Government.* Available at: http://www.constitution.org/jl/2ndtreat.htm

Loury, G. (1979). 'Market Structure and Innovation'. *Quarterly Journal of Economics*, 93: 395–410.

Machlup, F. (1958). *An Economic Review of the Patent System: Study No. 15.* Washington, DC: US Government Printing Office.

Mann, R.J. (2004). 'The Myth of the Software Patents Thicket: An Empirical Investigation of the Relationship between IP and Innovation in Software Firms'. Law and Economics Working Paper N. 022, February. Texas: The University of Texas School of Law.

Mansfield, E. (1986). 'Patents and Innovation: An Empirical Study'. *Management Science*, 32: 173–81.

Martin, C. (2004). 'Competitive Advantages through Cooperation in Patent Documentation: An Empirical Analysis of Oppositions in the Chemical Industry'. Unpublished diploma thesis, Munich School of Management, Munich.

Martinez, C. and Guellec, D. (forthcoming), 'Overview of Recent Changes and Comparison of Patent Regimes in the United States, Japan and Europe'. *Patents, Innovation and Economic Performance.* Paris: OECD.

Mathieu, A., Meyer, M., and van Pottelsberghe de la Potterie, B. (forthcoming). 'Turning Science into Business: A Case Study of a Major European Research University'. CEB Working Paper. Brussels: Université Libre de Bruxelles, Solvay Business School, Centre Emile Bernheim (CEB).

Matutes, C., Regibeau, P., and Rockett, K.E. (1996). 'Optimal Patent Protection and the Diffusion of Innovation'. *Rand Journal of Economics*, 27(1): 60–83.

Maurer, S.M. and Scotchmer, S. (2002). 'The Independent Invention Defense in Intellectual Property'. *Economica*, 69: 535–47.

Merges, R.P. and Nelson, R.R. (1990). 'On the Complex Economics of Patent Scope'. *Columbia Law Review*, 90: 839–916.

Mervin, M.J. and Warner, C.M. (1996). 'Techniques for Obtaining and Analyzing External License Agreements', in R.L. Parr and P.H. Sullivan (eds.) *Technology Licensing: Corporate Strategies for Maximizing Values.* New York: John Wiley & Sons, pp.187–205.

Meyer, M. (2002). 'Tracing Knowledge Flows in Innovation Systems'. *Scientometrics*, 54(2): 193–212.

—— (2003). 'Academic Patents as an Indicator of Useful Research? A New Approach to Measure Academic Inventiveness'. *Research Evaluation*, 12(1): 17–27.

Mgbeoji, I.L. (2003). 'The Juridical Origins of the International Patent System: Towards a Historiography of the Role of Patents in Industrialisation'. *Journal of the History of International Law*, pp.402–23.

Michel, J. (forthcoming). 'Considerations, Challenges and Methodologies for Implementing Best Practices in Patent Office and Like Patent Information Departments'. *World Patent Information.*

Miller, J. (2002). 'FESTO: Blessing to Patent Holders or Thorn in Their Sides?' *Duke Law & Technology Review*, 0017.

Milne, A.W. (1991). 'Very Broad Markush Claims: A Solution or a Problem? Proceedings of a Round-Table Discussion Held on August 29, 1990'. *Journal of Chemical Information and Computer Sciences*, 31(1): 9–30.

Moser, P. (2005). 'How Do Patent Laws Influence Innovation? Evidence from nineteenth-Century World's Fairs'. *American Economic Review*, 95(4): 1214–36.

Mossoff, A. (2001). 'Rethinking the Development of Patents: An Intellectual History, 1550–1800'. *Hasting Law Journal*, 52: 1255–65.

Mowery, D. and Sampat, B. (2004). 'The Bayh-Dole Act of 1980 and University–Industry Technology Transfer: A Model for Other OECD Governments?' *Journal of Technology Transfer*, 30(1–2): 115–27.

Mowery, D.C. and Ziedonis, A. (2002). 'Academic Patent Quality and Quantity Before and After the Bayh-Dole Act in the United States'. *Research Policy*, 31: 399–418.

—— Sampat, B., and Ziedonis, A. (2002). 'Learning to Patent: Institutional Experience, Learning, and the Characteristics of U.S. University Patents After the Bayh-Dole Act, 1981–1992'. *Management Sciences*, 48(1): 73–89.

Murray, F. and Stern, S. (2005). 'Do Formal Intellectual Property Rights Hinder the Free Flow of Scientific Knowledge? An Empirical Test of the Anti-Commons Hypothesis'. Working Paper, 11465. Cambridge, MA: NBER.

Narin, F., Hamilton, K., and Olivastro, D. (1997). 'The Increasing Linkage Between US Technology and Public Science'. *Research Policy*, 26(3): 317–30.

NAS (National Academy of Sciences) (2004). *A Patent System for the 21st Century*. Washington, DC: National Academy of Sciences.

Nelson, R.R. (2004). 'The Market Economy and the Scientific Commons'. *Research Policy*, 33(3): 455–71.

—— Sampat, B., Ziedonis, A., and Mowery, D. (eds.) (2004). *Ivory Tower and Industrial Innovation: University–Industry Technology Transfer Before and After the Bayh-Dole Act in the United States (Innovation and Technology in the World Economy)*. Palo Alto, CA: Stanford University Press, p.304.

Nickerson, J.A. (1996). 'Strategic Objectives Supported by Licensing', in R.L. Parr and P.H. Sullivan (eds.) *Technology Licensing: Corporate Strategies for Maximizing Values*. New York: John Wiley & Sons, pp.63–82.

Nordhaus, W.D. (1969). *Invention, Growth, and Welfare: A Theoretical Treatment of Technological Change*. Cambridge, MA: MIT Press.

Nozick, R. (1974). *Anarchy, State and Utopia*. New York: Basic Books.

Ochsenbein, P. (1987). 'The Patent Documentation Group (PDG)'. *World Patent Information*, 9: 92–5.

O'Donoghue, T. (1998). 'A Patentability Requirement for Sequential Innovation'. *Rand Journal of Economics*, Winter, 29(4): 654–79.

—— Scotchmer, S., and Thisse, J. (1998). 'Patent Breadth, Patent Life and the Pace of Technological Progress'. *Journal of Economics and Management Strategy*, 7(1): 1–32.

OECD (2002). *Genetic Inventions, Intellectual Property Rights and Licensing Practices*. Paris: OECD.

OECD (2004a). *Roundtable on Competition Policy*. Paris: OECD.

OECD (2004b). *Patents and Innovation: Trends and Policy Challenges*. Paris: OECD.

OECD (2005). *Background Report to the International Conference on IP as an Economic Asset: Key Issues in Exploitation and Valuation*. Berlin: OECD.

OECD (2006). *Compendium of Patents Statistics*. Paris: OECD.

O'Haver, R. (2003). 'Monetize your Intellectual Property'. *Mercer Management Journal*, 16: 60–6. Available at: http://www.mercermc.com/mmj

Ordover, J. (1991). 'A Patent System for Both Diffusion and Exclusion'. *Journal of Economic Perspectives*, 5(1): 43–60.

Pakes, A. and Simpson, M. (1989). 'Patent renewal data'. *Brookings Papers on Economic Activity, Microeconomics Annual*, pp.331–401.

—— —— (1991). 'The Analysis of Patent Renewal Data'. *Brookings Papers on Economic Activity, Microeconomic Annual*, pp.31–401.

Park, W. and Lippoldt, D. (2004). 'International Licensing and the Strengthening of IPR in Developing Countries'. Trade Policy Working Papers N. 10. Paris: OECD.

Parr, R.L. (1988). 'Fair Rates of Return'. *Patent World*, pp.36–41.

—— and Smith, G.V. (1994). 'Quantitative Methods of Valuing Intellectual Property'. in M. Simensky and L.G. Bryer (eds.) *The New Role of Intellectual Property in Commercial Transactions*. New York: John Wiley, pp.39–68.

Peeters, C. and van Pottelsberghe de la Potterie, B. (2006a). 'Innovation Strategy and the Patenting Behavior of Firms'. *Journal of Evolutionary Economics*, 16(1–2): 109–35.

—— —— (2006b). 'Understanding the Patenting Behaviour of Firms', in D. Bosworth and E. Webster (eds.) *The Management of Intellectual Property*. Northampton, MA: Edward Elgar.

Philipp, M. (2006). 'Patent Filing and Searching: Is Deflation in Quality the Inevitable Consequence of Hyperinflation in Quantity?' *World Patent Information*.

Pitkethly, R. (1997). 'The Valuation of Patents: A Review of Patent Valuation Methods with Consideration of Option Based Methods and the Potential for Further Research'. Working Paper. Oxford: Oxford Intellectual Property Research Centre.

Plant, A. (1934). *The Economic Theory Concerning Patents for Inventions Economica* (New Series) February, 1(1): 30–51.

Plogmann, V. (2005). 'IP Management in the Automobile Industry'. Paper presented at the OECD-EPO-BMWA Conference on IP as Economic Assets: Key Issues in Valuation and Exploitation, Berlin, June.

Prager, F. (1944). 'A History of Intellectual property From 1545 to 1787'. *Journal of the Patent Office Society*, 26(11).

Public Patent Foundation (2005). *Motion to the US Supreme Court in FTC v. Schering-Plough*. Available at: http://www.pubpat.org/PUBPAT_Schering_SCt_Brief.pdf

Putnam, J. (1996). 'The Value of International Patent Rights'. Unpublished PhD thesis, Yale University, New Haven, CT.

Quillen, C. and Webster, O. (2001). 'Continuing Patent Applications and Performance of the US Patent Office'. *Federal Circuit Bar Journal*, 11(1): 1–21.

—— —— and Eichmann, R. (2002). 'Continuing Patent Applications and Performance of the US Patent and Trademark Office'. *Extended Federal Circuit Bar Journal*, 12(1): 35–55.

Regibeau, P. and Rockett, K. (2004). 'The Relationship Between IP Law and Competition Law: An Economic Approach'. Mimeo, University of Essex.

Reitzig, M. (2004a). 'Strategic Management of Intellectual Property'. *MIT Sloan Management Review*, 45(3): 35–40.

—— (2004b). 'Improving Patent Valuations for Management Purposes–Validating New Indicators by Analyzing Application Rationales'. *Research Policy*, 33(6–7): 939–57.

Rolland Berger (2005). *Study on the Cost of Patenting in Europe*. Munich: EPO.

Romer, P. (1990). 'Endogenous technical change'. *Journal of Political Economy*, 94(5): 71–102.

—— (1991). 'Endogenous Technical Change'. Working Paper, 3210. Cambridge, MA: NBER.

Rowe, E.A. (forthcoming). 'The Experimental Use Exemption to Patent Infringement: Do Universities Deserve Special Treatment?' *Hastings Law Journal*, 57.

Rutz, B. and Yeats, S. (2004). 'The Importance of Being Inventive'. *EMBO Reports*, 5(2): 119–23.

Sakakibara, M. and Branstetter, L. (2001). 'Do Stronger Patents Induce More Innovation? Evidence from the 1988 Japanese Patent Law Reforms'. *Rand Journal of Economics*, 32(1): 77–100.

Sampat, B.N., Mowery, D.C., and Ziedonis, A. (2003). 'Changes in University Patent Quality After the Bayh-Dole Act: A Re-Examination'. *International Journal of Industrial Organization*, 21: 1371–90.

Samuelson, P. (1954). 'The Pure Theory of Public Expenditure'. *Review of Economics and Statistics*, 36(4): 387–9.

Sapsalis, E. and van Pottelsberghe de la Potterie, B. (2006). 'From Science to License: An Exploratory Analysis of the Value of Academic Patents'. Working Paper. Brussels: Université Libre de Bruxelles, Solvay Business School, Centre Emile Bernheim.

—— —— (forthcoming) 'The Institutional Sources of Knowledge and the Value of Academic Patents'. *Economics of Innovation and New Technology*.

—— —— Navon, R. (forthcoming) 'Academic Patenting vs. Industry Patenting: The Case of Biotechnology'. *Research Policy*.

Saragossi, S. and van Pottelsberghe de la Potterie, B. (2003). 'What Patent Data Reveal about Universities: The Case of Belgium'. *Journal of Technology Transfer*, 28: 9–15.

Schankerman, M. (1998). 'How Valuable is Patent Protection? Estimates by Technology Field'. *Rand Journal of Economics*, 29: 77–107.

—— Pakes, A. (1986). 'Estimates of the Value of Patent Rights in European Countries During the Post-1950 Period'. *Economic Journal*, 96: 1052–77.

Scherer, F.M. (1980). *Industrial Market Structure and Economic Performance*. Chicago: Rand McNally.

—— (1992). 'Schumpeter and Plausible Capitalism'. *Journal of Economic Literature*, 30(3): 1416–33.

—— (2002). 'The Economics of the Human Gene Patents'. *Academic Medicine*, 77(12 part 2): 1348–67.

Schmoch, U. (2004). 'The Technological Output of Scientific Institutions', in H. Moed, W. Glänzel and U. Schoch (eds.) *Handbook of Quantitative Science and Technology Research: The Use of Publication and Patent Statistics in Studies of Science and Technology Systems*. London: Kluwer Academic Publishers.

Schmookler, J. (1966). *Invention and Economic Growth*. Cambridge, MA: Harvard University Press.

Schumpeter, J.A. (1942). *Capitalism, Socialism and Democracy*. New York: Harper.

Scotchmer, S. (1999). 'On the Optimality of the Patent Renewal System'. *Rand Journal of Economics*, 30(2): 181–96.

—— (2004). *Innovation and Incentives*. Cambridge, MA: MIT Press, p.357.

—— Green, J. (1990). 'Novelty and Disclosure in Patent Law'. *Rand Journal of Economics*, 21(1): 131–46.

Shane, S. (2001). 'Technological opportunities and new firm creation'. *Management Science*, 47(2): 205–20.

Shapiro, C. (2001). 'Navigating the Patent Thicket: Cross-Licenses, Patent Pools, and Standard-Setting', in A. Jaffe, J. Lerner, and S. Stern (eds.) *Innovation Policy and the Economy*. Washington, DC: National Bureau of Economics.

Sheehan, J., Martinez, C., and Guellec, D. (forthcoming). 'Understanding Business Patenting and Licensing: Results of a Survey'. *Patents, Innovation and Economic Performance*. Paris: OECD.

Shepard, A. (1987). 'Licensing to Enhance Demand for New Technologies'. *Rand Journal of Economics*, 16: 360–8.

Silverman, S. (2001). *Einstein's Refrigerator and Other Stories from Flip Side Of*. Kansas City, MO: Andrews McMeel Publishing, p.192.

Simon, D. and Cohen, B. (2001). 'Complex Applications: A Return to First Principles'. *IIC: International Review of Industrial Property and Copyright Law*, 32(5): 485–606.

Sueur, T. (2004). 'How Do Third Party Patents Foster Innovation'. *Patents, Innovation and Economic Performance*. Paris: OECD.

Sullivan, S.Z. (1996). 'The Importance of Context in the Derivation of Royalty Rates', in R.L. Parr and P.H. Sullivan (eds.) *Technology Licensing: Corporate Strategies for Maximizing Values*. New York: John Wiley & Sons, pp.177–86.

Tandon, P. (1982). 'Optimal Patents with Compulsory Licensing'. *Journal of Political Economy*, 90: 470–86.

Thompson, M. and Rushing, F. (1999). 'An Empirical Analysis of the Impact of Patent Protection on Economic Growth: An Extension'. *Journal of Economic Development*, 24(1): 67–76.

Tietze, G., Granstrand, H., and Herstatt, C. (2006). 'Towards Advanced Intellectual Property Management: Events and Stages During the Development—Evidence from the Biotech Sector'. Working Paper No. 35. Hamburg: Technologie und Innovations management, Technische Universität Hamburg-Harburg.

Tong, X. and Frame, J.D. (1994). 'Measuring National Technological Performance with Patent Claims Data'. *Research Policy*, 23(2): 133–41.

Trajtenberg, M. (1990). 'A Penny for Your Quotes: Patent Citations and the Value of Innovations'. *Rand Journal of Economics*, 21(1): 172–87.

Tufano, P. (1989). 'Financial innovation and first mover advantages.' *Journal of Financial Economics*, 25: 213–40.

UK High Court (2005). Case [2005] EWHC 1589 (Pat), Judge Prescott.

Van Looy, B., Callaert, J., and Debackere, K. (2006a). *Publication and Patent Behaviour of Academic Researchers: Conflicting, Reinforcing or Merely Co-existing?* Available at: http://ssrn.com/abstract=874905

—— Du Plessis, M., and Magerman, T. (2006b). 'Data Production Methods for Harmonized Patent Indicators: Assignee sector allocation'. Working Paper and Studies. Luxemburg: EUROSTAT.

Van Ophem, H., Brouwer, E., Kleinknecht, A., and Mohnen, P. (2001). 'The Mutual Relation between Patents and R&D', in A. Kleinknecht and P. Mohnen (eds.) *Innovation and Firm Performance: Econometric Explorations of Survey Data*. New York: Palgrave.

Van Pottelsberghe de la Potterie, B. and François, D. (2006). 'The Cost Factor in Patent Systems'. EPO Working Papers CEB 06-002. RS. Brussels: Université Libre de Bruxelles, Solvay Business School Centre Emile Bernheim(CEB).

Von Scheffer, G. and Zieger, M. (2005). 'Methods for Patent Valuation'. Paper presented at the International Conference on IP as an Economic Asset: Key Issues in Exploitation and Valuation, Berlin.

Van Zeebroeck, N., van Pottelsberghe de la Potterie, B., and Guellec, D. (2006). 'Claiming More: The Increased Voluminosity of Patent Applications and its Determinants'. EPO Working Paper CEB 06-018. RS. Brussels: Université Libre de Bruxelles, Solvay Business Centre, Centre Emile Bernheim (CEB).

Walsh, J.P., Arora, A., and Cohen, W.M. (2002). 'The Patenting and Licensing of Research Tools and Biomedical Innovation'. *Science, Technology and Economic Policy Board of the National Academies*. Washington, DC: National Academies Press.

———— ———— ———— (2003). 'Effects of Research Tool Patents and Licensing on Biomedical Innovation', in W.M. Cohen and S.A. Merrill (eds.) *Patents in the knowledge-based Economy*. Washington, DC: National Academies Press.

———— ———— ———— (2006). 'Roadblocks to Accessing Biomedical Research Tools'. Paper presented at the CSIC/OECD/OEPM Conference on the Research Use of Patented Inventions. Madrid, Spain, 18–19 May.

Weitzman, M. (1998) 'Recombinant Growth'. *Quarterly Journal of Economics*, May, CXIII: 331–60.

Wright, B. (1983). 'The Economics of Invention Incentives: Patents, Prizes, and Research Contracts'. *The American Economic Review*, 73(4): 691–707.

Yeats, S. (2005). 'Patenting Human Genes and Stem Cells'. *The Ethics of Patenting Human Genes and Stem Cells*. Copenhagen: Danish Council of Ethics. Available at: http://www1.etiskraad. dk/graphics/03_udgivelser/engelske_publikationer/patenting_human_genes/patents_konf_0 4/kap05.htm

Zucker, L.G. and Darby, M.R. (1996). 'Star Scientists and Institutional Transformation: Patterns of Invention and Innovation in the Formation of the Biotechnology Industry'. *Proceedings of the National Academy of Science*, 93: 12709–16.

■ INDEX